Annals of Mathematics Studies

Number 121

Cosmology in (2+1)-Dimensions, Cyclic Models, and Deformations of $M_{2,1}$

by

Victor Guillemin

PRINCETON UNIVERSITY PRESS

PRINCETON, NEW JERSEY

1989

Printed in the United States of America
by Princeton University Press, 41 William Street
Princeton, New Jersey

Library of Congress Cataloging-in-Publication Data

Guillemin, V., 1937-
    Cosmology in (2+1)-dimensions, cyclic models and
deformations of M2,1.

    (Annals of mathematics studies; no. 121)
    Bibliography: p.
    1. Cosmology–Mathematical models. 2. Geometry,
Differential. 3. Lorentz transformations. I. Title.
II. Series.
QB981.G875 1988   523.1'072'4   88-19517
ISBN 0-691-08513-7 (alk. paper)
ISBN 0-691-08514-5 (pbk.)

CONTENTS

Cosmology in (2+1)-Dimensions,

Cyclic Models, and Deformations of $M_{2,1}$

FOREWORD

In this paper a "cyclic model" will mean a compact Lorentz manifold with the property that all its null-geodesics are periodic. Such a model is cyclic in the sense that every space-time event gets replicated infinitely often; it has an infinite number of antecedents with identical "pasts" and "futures". We should warn the non-expert that this is not what relativists usually mean by cyclicity. This term is almost always used to describe periodic solutions of Einstein's equations. In (2+1)-dimensions this implies that the metric involved is conformally flat; and, as we will see in §11, this is practically incompatible with cyclicity in our sense.

We will call a Lorentz metric all of whose null-geodesics are periodic a Zollfrei metric. (For the etymology of this term, see §1.) Notice that the property of being Zollfrei is conformally invariant. This is because two Lorentz metrics have the same null-geodesics if they differ by a conformality factor. (Another way of stating this fact is that the trajectories of light rays are independent of the metric structure of space-time but only depend on its causal structure: i.e., the specification of the future of every space-time event.)

The Zollfrei problem is interesting even in dimension 2;
in fact, as a warm-up for the problem in dimension 3 we will
briefly describe what happens in dimension 2:

<u>Theorem</u>. Let g_{can} be the standard Zollfrei metric on
$S^1 \times S^1$; i.e., the metric, $d\theta_1 \, d\theta_2$, where θ_1 and θ_2 are the
standard angle variables on the first and second factors. Let
(X, g) be any oriented Zollfrei two-fold. Then there exists a
covering map $\pi: X \to S^1 \times S^1$ such that $\pi^* g_{can}$ and g are
conformally equivalent

<u>Proof</u>: First of all notice that every oriented compact
Lorentzian two-fold has to be diffeomorphic to $S^1 \times S^1$ since
its Euler characteristic is zero. Now suppose that X is a
compact Lorentzian two-fold all of whose null-geodesics are
periodic. The null-cone at $p \in X$ consists of two lines in
T_p (See figure.)

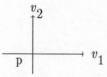

so the conformal geometry of X is completely described by a
pair of tranverse line element fields. Let v_1 and v_2 be
vector fields defining these line element fields. By assumption the integral curves of v_1 and v_2 are all closed.

Choose an oriented curve, γ_1, which intersects each integral curve of v_1 transversally. (This is always possible. See [41], page 9.) Let p be the intersection number of γ_1 with the trajectory of v_1 through x. It is clear that this number is independent of x. Thus γ_1 has to intersect each trajectory in *exactly* p points since the orientation numbers at the points of intersect have to be all of the same sign. (See figure.)

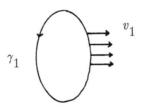

Suppose in particular that x is on γ_1. Let f(x) be the next point at which the trajectory through x intersects γ_1. The map f: $\gamma_1 \to \gamma_1$ which sends x to f(x) is a diffeo— morphism of γ_1, and the points, x, f(x), $f^2(x),\ldots,f^{p-1}(x)$, are the distinct points where the trajectory through x inter— sects γ_1. Thus f defines a *free* action of the finite cyclic group, Z_p, on γ_1. In particular there exists a covering map

$$\phi_1: \gamma_1 \longrightarrow S^1$$

whose fibers are the Z_p orbits. Now extend ϕ_1 to all of X by associating to the point $x \in X$ the Z_p orbit in which the

trajectory through x intersects γ_1. Notice that the level
curves of ϕ_1 are identical with the integral curves of v_1.

Next choose a cycle, γ_2, in X which intersects each
integral curves of v_2 transversally and repeat this argument.
Let $\phi_2: X \to S^1$ be the analogue of the mapping, ϕ_1, for the
v_2 trajectories, and let

$$\phi: X \to S^1 \times S^1$$

be the product of ϕ_1 and ϕ_2. We leave it for the reader to
convince himself that ϕ is a covering map and that $\phi^* g_{can}$
is conformally equivalent to g.   Q.E.D.

An easy corollary of this theorem is that every oriented
Zollfrei two-fold is of the form $\mathbb{R}^2/L$, L being a rational
lattice subgroup of $\mathbb{R}^2$ and the null-geodesics being the
projections of the lines parallel to the x and y axes.

Lets next turn to the Zollfrei problem in dimension
three. We pointed out above that for a compact oriented
two-manifold to be a Lorentz manifold it has to be diffeomorphic to T^2. Unfortunately the fact that a compact
3-manifold, M, is a Lorentz manifold is no constraint at all
on the topology of M. (The only topological obstruction to
the existence of a Lorentz structure on M is the vanishing of
its Euler characteristic, which is automatic in dimension

three.) We suspect, however, that for M to be Zollfrei it
must have a very simple topological structure. To be more
specific, Thurston's classification of three-dimensional
geometries suggests four obvious possibilities for the diffeo—
type of M and we suspect these are the only possibilities.
We recall that a geometry for Thurston is a simply connected
homogeneous space of the form G/H where G is a connected
Lie group and H a compact subgroup. Thurston calls a compact
manifold geometrizable if it is universal cover is such a
space. He has conjectured that all three-manifolds can be
obtained from the geometrizable ones by simple topological
operations like "connected sum." The three-dimensional
geometries are easy to classify and turn out to be eight in
number; so every geometrizable three-manifold belongs to one of
eight distinct categories. Our conjecture is tha the Zollfrei
examples are all geometrizable and belong to the simplest of
these eight categories, namely $S^2 \times \mathbb{R}$, with structure group,
$G = SO(3) \times \mathbb{R}$. The compact manifolds with $S^2 \times \mathbb{R}$ as universal
cover are just four in number: $S^2 \times S^1$ and the three spaces
with $S^2 \times S^1$ as double cover corresponding to the three
involutions of $S^2 \times S^1$:

$i)$ $\qquad\qquad\qquad (x,y,z,t) \longrightarrow (-x,-y,-z,t)$

$ii)$ $\qquad\qquad\qquad (x,y,z,t) \longrightarrow (-x,-y,-z,t+\pi)$

$iii)$ $\qquad\qquad\qquad (x,y,z,t) \longrightarrow (-x,-y,z,t+\pi).$

(See [34], page 458). All of these spaces have Zollfrei
metrics which are covered by the standard Einstein metric,
$(dx)^2 + (dy)^2 + (dz)^2 - (dt)^2$, on $S^2 \times \mathbb{R}$. We will henceforth
call metrics of this type standard Zollfrei metrics; so our
conjecture can be reformulated in the form:

Conjecture. Every Zollfrei manifold in dimension three has the
same diffeotype as one of the standard examples.

A somewhat safer conjecture is that this conclusion is
true with the additional hypothesis that the universal cover of
M satisfies the causality condition (i.e., has no closed
space-like or time-like curves. See [29], page 407. Incidentally, for the Floquet theory, which we will describe below,
this property is highly desirable.)

If the above conjecture were true, the natural place to
look for Zollfrei metrics in dimension three would be in the
vicinity of the standard models. In fact an obvious question
to ask is: Do the standard models admit non-trivial "Zollfrei
deformations"?

This question will occupy us for the next 150 pages. We
will, for the most part, concentrate on the simplest and most
symmetric standard model, the conformal compactification of
Minkowski three space, which has the toplogy of the second
space on the list above. We will henceforth denote this model

by $M_{2,1}$. (See §2.) We will show in §12 that it has lots of
non-trivial Zollfrei deformations.

It would be interesting to develop a deformation theory
along the lines of this monograph for some of the other
standard models; and, in fact, we hope to do so sometime in the
future. Particularly intriguing is the third space on the
list, above. This space is diffeomorphic to the connected sum
of $\mathbb{RP}^3$ with itself and is the only geometrizable three-
manifold which is a connected sum ([34], page 457. It is quite
a challenge, by the way, to describe the Zollfrei metric of
$\mathbb{RP}^3 \# \mathbb{RP}^3$ as a "connected sum" of metrics on the individual
$\mathbb{RP}^3$'s.)

There are interesting cyclicity phenomena in dimension
three which we unfortunately won't have time to pursue in this
article. We will, however, briefly describe the most bizarre
of these: One of the eight geometries of Thurston is the
universal cover of $SL(2,\mathbb{R})$. (This geometry plays an important
role in the study of Seifert fiber spaces. See [24].) From
the Killing form on the Lie algebra of $SL(2,\mathbb{R})$ one gets a
bi-invariant Lorentz metric on $SL(2,\mathbb{R})$ with the property that
all time-like geodesics are periodic. The null-geodesics on
the other hand are not periodic. Compact examples of this
phenomenon can be obtained by quotienting $SL(2,\mathbb{R})$ by a
discrete co-compact subgroup. (See [24] for details.)

In this article we won't, for the most part, discuss Zollfrei metrics in dimension greater than three. This is not because this problem is uninteresting but because it seems to be much harder to find examples. (Some examples do exist: compactified Minkowski n-space, $S^n \times S^1$, $\mathbb{CP}^n \times S^1$ etc. More generally if M is any of the n-dimensional $SO(n)$-invariant Zoll manifolds constructed by Weinstein in [2], $M \times S^1$ is an example.) It is not unlikely that other methods than ours (for instance twistorial methods) will yield a larger supply of examples.

Having given some indication of the contents of this monograph we will say a few words about our motives for writing it. One of our main motives is the flickering (and probably unwarranted) hope that there are interesting solutions of Einstein's equations in (3+1) dimensions associated with the cyclic models described above. More explicitly the standard $M_{2,1}$ is what is left of the anti-deSitler universe after it undergoes gravitational collapse. We suspect that there may be interesting solutions of Einstein's equations in (3+1) dimensions which are related in the same way to cyclic deformations of $M_{2,1}$. The evidence for this is unfortunately still rather skimpy: First of all, as we will see in §12, the solutions of the linearized Einstein equations on $M_{3,1}$ (aka "free gravitons") are in one-one correspondence with the infinitesmal cyclic deformations of $M_{2,1}$. Secondly, there are methods, developed independently by Lebrun and by Fefferman and Graham, for putting (3+1)-dimensional Einstein "collars" on (2+1)-dimensional conformal spaces, which are particularly well adapted to (2+1)-dimensional cyclic models in our sense. (See the comments at the end of §15.)

Our second motive for writing this monograph is more defensible, at least on mathematical ground. Namely, as we will see below, Zollfrei manifolds turn out to have lots of interesting non-local conformal invariants. To construct these invariants we make use of some ideas of Paneitz and Segal of which we will give a short description here. (More details will be provided in §18.) Let M be a compact manifold, $\Box$ a differential operator on M and $\tilde{M}$ the universal cover of M. Corresponding to $\Box$ is a differential operator on $\tilde{M}$ which we will denote by $\tilde{\Box}$. It is clear that the action of the fundamental group of M on $\tilde{M}$ leaves $\tilde{\Box}$ fixed; hence there is a canonical representation of the fundamental group of M on the space of solutions of the equation $\tilde{\Box} = 0$. A classical example of this situation is Hill's equation

$$\frac{d}{dt^2} + q(t) = \Box = 0$$

on the circle (i.e. $q(t) = q(t + 2\pi)$.) The deck transformation, $\sigma: t \to t+2\pi$, acts on the two dimensional space of solutions of $\tilde{\Box} = 0$ and the two-by-two matrix, $A(\sigma)$, representing

σ is the classical Floquet matrix. Paneitz and Segal point out that if M is a compact space—time and □ is a conformally invariant differential operator, the space of solutions of $\tilde{\Box}$ will very often be a fairly manageable space, (e.g. Hilbertizable). They also show that the Floquet representation on this space describes what is traditionally referred to as the "scattering phenomena" associated with $\tilde{\Box}$. (See [30], [31], [21] and [37].)

We will review the theory of these Floquet operators in part five and will show that if M is Zollfrei they have the form

$$e^{i\sigma}I + K.$$

Here σ is an integral multiple of $\pi/4$ and K is a compact operator. (Incidentally we will also show that the converse of this assertion is true. If the Floquet operators are of this simple form M has to be Zollfrei.) In particular for Zollfrei manifolds these operators have discrete spectrum; so these manifolds have a large number of discrete conformal invariants.

It would be nice to relate these invariants to other conformal invariants of M, for instance the Chern-Simon invariant [5], or the invariants studied by Fefferman-Graham [8] and Branson and Oersted [3]. At first glance, however, they seem

to be a good deal more complicated. (We will describe some of
our efforts to compute these invariants in §19.)

We will conclude this introduction by warning the reader
about some of the technical complications involved in dealing
with closed geodesics on Lorentzian manifolds. As was pointed
out to us by John Beem a closed null-geodesic does not need to
be complete. The problem is that the following situation can
occur. (In fact it can occur generically. See [1].) Let H be
the Hamiltonian function on T*M defining geodesic flow and
(p,ξ) a point on the null-energy surface H = 0. A trajectory
of geodesic flow whose initial point is (p,ξ) can return
after a finite period of time, T, not to the point (p,ξ)
itself but to the point $(p,\lambda\xi)$ with $\lambda > 1$. The projection
of this trajectory onto M will look like a perfectly
respectable closed null-geodesic. Notice, however, that the
next circuit which this trajectory makes will go from $(p,\lambda\xi)$
to $(p,\lambda^2\xi)$ in time T/λ and the next circuit after that from
$(p,\lambda^2\xi)$ to (p,λ^3) in time T/λ^2. The ultimate destiny of
this trajectory is clear: It will cease to exist after a
finite period of time.

To avoid this kind of behavior we will categorically
decree from now on that Zollfrei ⇒ the trajectories of geodesic
flow are periodic on the null-energy surface H = 0.

There is another type of pathology which is not quite as
serious as this but which we will rule out to make life simpler

for ourselves. Namely, it is possible for all trajectories of
geodesic flow on H = 0 to be periodic, but with some trajectories much shorter than the rest. This situation is illus-
trated in the figure below:

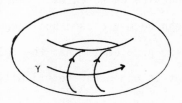

(γ is the short geodesic and all the others spiral around it
giving rise to a Seifert fibration of the solid torus).

To avoid this type of behavior we will decree that
Zollfrei $\Rightarrow$ geodesic flow is a fibration (in the usual sense) of
the energy surface, H = 0, by S^1's.

Before we get down to business we would like to express
our appreciation to the many persons who have helped us with
the preparation of this manuscript. The material on the
infinitesmal deformations of compactified Minkowski space could
not have been written without the help of David Vogan. (Our
original version of this material, using spherical harmonics,
was three times as long as the Harish-Chandra module approach
described in §9-12.) Similarly the sections on the microlocal
properties of the x-ray transform,, §14—17, are much better in
the final manuscript than they were in their original version

thanks to Richard Melrose's unstinting aid.   Other persons with whom we discussed the contents of this manuscript and who provided us with valuable suggestions for improving it are John Beem, Luis Casian, Michael Eastwood, John Morgan, Bent Oersted, Michele Vergne,, Gunther Uhlmann and Alejandro Uribe.   Last but not least, the stimulus for writing this paper was Irving Segal's monograph, [35] (which convinced us that the Einstein static universe still has to be taken seriously as a cosmological model.)

PART I

A RELATIVISTIC APPROACH TO ZOLL PHENOMENA

§1.   A Riemannian metric on a compact manifold is called a
Zoll metric if all of its geodesics are simply periodic of
period 2π. For instance the standard metric, $(dx)^2$, on S^2
has this property. Seventy—five years ago, Funk wrote a
seminal paper on Zoll two—folds [9] in which he posed the
following problem: find all Zoll metrics on S^2 which are C^2
close to the standard one. In particular he proposed an
algorithm for constructing such metrics: Given a function f
on S^2 define the Funk transform, $\hat{f}$ of f to be the
function

$$\hat{f}(p) = \int_{\gamma_p} f ds ,$$

where γ_p is the hemispherical circle on S^2 obtained by
situating p at the north pole. It is easy to see that $\tilde{f} \equiv 0$
if and only if f is odd, i.e. $f(-p) = -f(p)$ for all
$p \in S^2$. Starting with an odd function, f_0, Funk shows how to
construct a sequence of functions, $f_1, f_2, \ldots$, by solving
recursively integral equations of the form

16

$$\hat{f}_i = F_i(f_1,\ldots,f_{i-1}),$$

and conjectures, first of all, that the series

(*) $$f_t = \sum f_i(x) t^i$$

converges for all t in a sufficiently small interval about $t = 0$, and, secondly, that $(1+f_t)(dx)^2$ is a Zoll metric for all t on this interval.

Given the convergence of (*) the second assertion is quite plausible; but it is not known to this day whether (*) converges except for very special choices of f_0. (In fact it seems unlikely that it does.) There is, however, an updated version of the Funk algorithm involving Nash—Moser techniques, for which (*) does converge, and that gives essentially the same result as that which Funk had hoped to get from the scheme above. (For a survey of what is known about Zoll metrics and the Funk problem, see [2].)

Now suppose we are given a Riemann metric, $(d\gamma)^2$, on S^2. Let t be the standard angle variable on S^1 and consider the pseudometric of signature $(2+1)$:

(1.1) $$(d\gamma)^2 - (dt)^2,$$

on $S^2 \times S^1$. It is easy to see that the metric $(d\gamma)^2$ is a Zoll metric if and only if all of the null-geodesics of (1.1) are simply periodic. This suggests a generalization of the Funk problem: find all pseudometrics of Lorentzian type on $S^2 \times S^1$ which are C^2 close to the product metric

$$(1.2) \qquad\qquad (dx)^2 - (dt)^2$$

and have the property described above. To distinguish these metrics from the usual Zoll metrics (and at the same time call attention to their similarities) we will henceforth call these metrics Zollfrei metrics.

Notice, by the way, that the metrics, (1.1), are *isochronous*, i.e. invariant with respect to the group, S^1, of displacements in time. A sub-problem which is interesting in its own right is to determine all Zollfrei metrics on $S^2 \times S^1$, close to the standard one, which are isochronous. It turns out, as we will show later, that there are metrics of this type which are *not* conformally conjugate to a metric of the form (1.1).

§2.   There is a second variant of the Funk problem which is also interesting: Recall that ordinary Minkowski three-space possesses a conformal involution

(2.1) $x \longrightarrow (-x_0^2 + x_1^2 + x_2^2)^{-1} x$ . .

This involution is not globally defined, but it becomes so if we attach to $\mathbb{R}^3$ a light cone at infinity and require (2.1) to interchange the light cone at infinity with the light cone at zero. The manifold obtained by this construction looks as in the figure below. See [18].

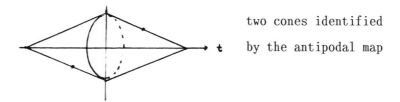

two cones identified by the antipodal map

In this figure $(x, t+\pi) \sim (-x, t)$. It is compact and (as is clear from the way it was constructed) possesses an intrinsic conformal (or causal) structure. We will henceforth refer to it as *compactified Minkowski space* and denote it by $M_{2,1}$.

Here are some other well—known descriptions of $M_{2,1}$

I. (due to Veblen) Consider the quadratic form

$$Q(x) = -x_1^2 - x_2^2 + x_3^2 + x_4^2 + x_5^2$$

on $\mathbb{R}^5$. $M_{2,1}$ is the projective quadric in $\mathbb{R}P^4$ defined by
the equation $Q = 0$. Such a quadric possesses an intrinsic
$SO(3,2)$-invariant conformal structure of Lorentzian type. The
null-geodesics are just the null-lines; and, since each null
line intersects the hyperplane, $x_1 = 0$, in one point, the set
of null—lines forms an $\mathbb{R}P^3$. For a discussion of this model,
see [35].

Since the double cover of $\mathbb{R}P^4$ is the four-sphere,
$x_1^2 + \cdots + x_5^2 = 2$, it is clear that the double cover of $M_{2,1}$ is
the set of points in $\mathbb{R}^5$ satisfying

$$x_1^2 + \cdots + x_5^2 = 2 \qquad \text{and} \qquad -x_1^2 - x_2^2 + x_3^2 + \cdots + x_5^2 = 0$$

or, separating variables,

$$x_1^2 + x_2^2 = 1 \qquad \text{and} \qquad x_3^2 + x_4^2 + x_5^2 = 1$$

i.e. it is just $S^2 \times S^1$. It is easy to see that the metric
(1.2) is compatible with the canonical conformal structure
mentioned above. In other words, *the Einstein universe is
causally isomorphic to the double cover of* $M_{2,1}$.

II. (due to Klein.) Let ω be the standard symplectic two-form on $\mathbb{R}^4$ and let $Sp(2,\mathbb{R})$ be the group of linear mappings of $\mathbb{R}^4$ which preserve ω. Once can show that $Sp(2,\mathbb{R})$ is the double cover of $SO(3,2)$, and this fact implies that $M_{2,1}$ is the Grassmannian of Lagrangian subspaces of $\mathbb{R}^4$. This description of $M_{2,1}$ has the virtue of providing one with an interesting complexification of $M_{2,1}$. Namely, let $\omega_{\mathbb{C}}$ be the $\mathbb{C}$–linear extension of ω to $\mathbb{C}^4$, and let $(M_{2,1})_{\mathbb{C}}$ be the set of all Lagrangian subspaces of $\mathbb{C}_4$. Contained in $(M_{2,1})_{\mathbb{C}}$ is the set, $M_{2,1}^{+}$, of all positive–definite Lagrangian subspaces of $\mathbb{C}^4$. This set turns out to be the Siegel domain, H_3, and its Shilov boundary is $M_{2,1}$.*

Since the property of being a Zollfrei metric is a conformal or causal property, it is natural to pose the problem of existence of Zollfrei structures not only for $S^2 \times S^1$ but for $M_{2,1}$ as well. The $M_{2,1}$ problem, incidentally, has markedly different features from the $S^2 \times S^1$ problem. For instance there are no isochronous Zollfrei deformations of the standard causal structure on $M_{2,1}$, but there are lots of such deformations of $S^2 \times S^1$. We will also see that the existence of Zollfrei deformations of $M_{2,1}$ is closely related to the fact that $M_{2,1}$ is the Shilov boundary of the Siegel domain, H_3.

---

*H_3 is usually defined to be the set of complex symmetric 2x2 matrices, A+iB, with B postive definite.

One last comment of a general nature: What about the existence of Zollfrei metrics on $S^{n-1} \times S^1$ or $M_{n-1,1}$ for $n > 3$? At the moment we know of some examples of Zollfrei metrics on $S^{n-1} \times S^1$ but of *no* such metrics on $M_{n-1,1}$ except for the standard one. It is an interesting question whether the standard Zollfrei metric on $M_{n-1,1}$ is *rigid*. This is not entirely trivial since $M_{n-1,1}$ admits lots of infinitesmal deformations. However, if $n > 3$ there are formal obstructions to finding deformations corresponding to them. The first of these is the quadratic obstruction, [23], of Kiyohara. We suspect that it is impossible for an infinitesmal deformation of $M_{n-1,1}$ to get past this quadratic obstruction.

§3.   Here is a brief summary of the contents of the paper: For simplicity we will concentrate on the *deformation* question for Zollfrei metrics.   This can be conveniently subdivided into two parts:

I: Determine all infinitesmal Zollfrei deformations of the standard conformal structure on $M_{2,1}$ (or on $S^2 \times S^1$).

II: Can every infinitesmal deformation be extended to a deformation?

In the second part of the paper we will concentrate on question II.   In particular we will show:

1.   There are no *formal* obstructions to extending an infinitesmal deformation to a deformation.

2.   The question of whether every infinitesmal deformation is extendible to a deformation can be reduced to apriori estimates for a certain "generalized x-ray transform".   (See below.)

In the third part of this paper we will concentrate on question I.   We will show, for instance, that there are no infinitesmal isochronous deformations of $M_{2,1}$ and that $S^2 \times S^1$ possesses, in addition to the isochronous deformations of type (1.1) (which we will call "deformations of lapse type") another class of isochronous deformations (which we will call of

"shift" type.) Our most surprising result is that the
infinitesmal deformations of $M_{2,1}$ can be thought of as "free
gravitons". More explicitly we will show that the infinitesmal
deformations of $M_{2,1}$ can be identified with sections of a
holomorphic vector bundle over the Siegel domain, $M_{2,1}^{+}$.
Unfortunately, not all sections of this vector bundle correspond to infinitesmal deformations. We will show that those
that do have to satisfy a rather mysterious-looking first order
differential equation. It turns out, however, that this
equation has a simple explanation in terms of structures living
on four-dimensional compactified Minkowski space. Namely $M_{3,1}$
is the Shilov boundary of the four-dimensional Siegel domain,
$M_{3,1}^{+}$, and $M_{2,1}^{+}$ sits inside $M_{3,1}^{+}$ as a globally-Cauchy
hypersurface. The mysterious equation we've just referred to
turns out to be the compatibility conditions satisfied by
Cauchy data for solutions of the mass-zero, spin-two field
equations on $M_{3,1}^{+}$!

The fourth part of this paper will be devoted to some
questions in integral geometry: Given a Zollfrei metric,
$(d\sigma)^{2}$, on $M_{2,1}$ and its manifold, P, of null geodesics,
consider the integral transform

$$(3.1) \qquad\qquad R(d\sigma)^{2}\colon \Omega^{1}(M_{2,1}) \longrightarrow C^{\infty}(P)$$

defined by the formula

$$(3.2) \qquad\qquad (R\omega)(\gamma) = \int_\gamma \omega$$

at $\gamma \in P$. If $(d\sigma)^2$ is the standard Einstein metric, (3.1),
is a relativistically invariant version of the usual x-ray
transform (see [4]); so we will refer to (3.1) as the "generalized x-ray transform." In part 4 we will study the microlocal structure of this transform. In particular we will show
that it is a Fourier integral operator in Hörmander's sense and
that it involves a canonical relation of the form.

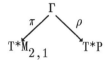

where π is a fold map and ρ is a "blowing–down" of Γ along
the fold set of π. In §17 we will construct a parametrix for
R^t, and use it to establish the estimates alluded to above. A
key ingredient in this construction is a "blowing-up" argument
for which we are indebted to Richard Melrose. (A remark on the
prerequisites for reading this part of the article: Almost all
the materials on F.I.O.'s is now available in Hörmander's
beautiful book, [20]. However, to make the arguments in §16-17
completely self-contained we would have been forced to include
some results on paired Lagrangian distributions which are not
in Hörmander. We decided instead to omit these results

altogether. They will be published elsewhere. The material on
integral geometry is reviewed, in a slap-dash way, in §14; but
it would be helpful for the reader to have some prior acquain-
tance with the so-called "Gelfand-Helgason" side of integral
geometry. The review articles by ourselves, by Gindinkin and
by Helgason in [4] are a good source for this material. For
those who intend to skip this part entirely a brief glance at
§16, appendix A is advisable.)

In the last part of this paper we will look briefly at
spectral properties of Zollfrei manifolds. We have already
said a few words in the introduction about the Floquet theory
of Segal and Paneitz. The Floquet invariants of Zollfrei
manifolds are very closely related to the band invariants of
Zoll manifolds and, like these invariants, are quite hard to
compute. A particularly challenging invariant of this sort is
the *determinant* of the Floquet operator. We will report on
what we know about it in §19.

PART II

THE GENERAL THEORY OF ZOLLFREI DEFORMATIONS

§4. We will show in §7 how to formulate the deformation problem for Zollfrei metrics as an existence problem for a functional equation in an appropriate Frechet space. To do this we will be forced to consider a slightly more general class of causal structures on $M_{2,1}$ than the class of causal structures associated with Lorentzian pseudo-metrics. What we will need is a "Finslerian" notion of causal structure of the type introduced by Segal in [35]: Let M be a manifold, p a fixed point of M and T_p the tangent space to M at p. We will define a *causal structure* on T_p to be an open convex cone, $C_p \subseteq T_p$ such that $C_p \cap (-C_p)$ is empty. We will denote by C_p^* the dual cone in T_p^*. We will say that C_p is smooth (strictly convex) if for every $\xi \in C_p^*$ and every $a > 0$ the intersection of C_p with the hyperplane

$$\{v \in T_p, \ \langle v, \xi \rangle = a\}$$

is smooth (strictly convex). From now on we will assume, without saying so, that *all* the cones we consider have these two properties (though this excludes some interesting examples of causal structures. See [35], Chapter 2.)

By a *causal structure* on a manifold M we mean the smooth assignment of a convex cone, C_p, to each point, p, in M. By this we mean that the set

$$C = \bigcup_{p \in M} C_p$$

is a smooth, open fiber subbundle of the punctured tangent bundle of M.

We will call the boundary of C_p in $T_p - 0$, i.e. ∂C_p, the forward light cone in T_p and $K_p = \partial C_p \cup (-\partial C_p)$ the (full) light cone in T_p. We will also define in the same way the dual objects in T_p^*:

$$\Sigma_p^+ = \partial C_p^* \qquad \text{and} \qquad \Sigma_p = \partial C_p^* \cup (-\partial C_p^*).$$

By assumption Σ_p varies smoothly as one varies p, so the subset

$$\Sigma = \bigcup_{p \in M} \Sigma_p$$

of the punctured cotangent bundle of M is actually a smooth subbundle. We will call a function, H, on T^*M-0 a *defining function* of Σ if

$$H(p,\xi) = 0 \iff \xi \in \Sigma_p$$

(4.1) and

$$\frac{\partial H}{\partial \xi_i}(p,\xi) \neq 0 \quad \text{for some} \quad i \quad (\text{depending on} \quad \xi) \quad \text{if} \quad \xi \in \Sigma_p.$$

Without loss of generality we can assume that $H(p,\xi) > 0$ when $\xi \in C_p$ and that for all $\xi \in T^*_p\text{-}0$ and all $\lambda \neq 0$

(4.2) $$H(p,\lambda\xi) = \lambda^2 H(p,\xi).$$

Let Ξ_H be the Hamiltonian vector field associated with H. For any point $(p,\xi) \in \Sigma$ the integral curve of Ξ_H through (p,ξ) lies on Σ. These integral curves depend, of course, on the choice of H, but it is easy to see that they depend on H only up to choice of a parametrization. In fact, any other fucntion, H_1, with the properties (4.1) and (4.2) has to be of the form, $H_1 = FH$ for some positive C^∞ function, F, which is homogeneous of degree zero in ξ. (And hence *bounded* on all compact subsets of M.) Notice now that

$$\Xi_{H_1} = H\Xi_F + F\Xi_H \ .$$

On Σ the first term vanishes and

(4.3) $$\Xi_{H_1} = F\Xi_H \ ,$$

so, except for reparametrization, Ξ_{H_1} and Ξ_H have the same integral curves.

At each point, p, of M the restriction of H to T_p^* defines a legendre transform

$$(4.4) \quad \mathcal{L}_H \colon T_p^*\text{-}0 \to T_p, \quad \mathcal{L}_H(\xi) = \left[\frac{\partial H}{\partial \xi_i}(p,\xi),\ldots,\frac{\partial H}{\partial \xi_n}(p,\xi)\right].$$

Because of the fact that C_p is strictly convex, (4.4) maps the forward light cone, Σ_p^+, in $T_p^*\text{-}0$ bijectively onto the light cone ∂C_p in $T_p\text{-}0$. Suppose now that one is given a point, p, in M and a light vector $v \in \partial C_p$. Let ξ be the point on Σ_p^+ corresponding to v and let $\gamma(t)$, $-a < t < a$ be the integral curve of Ξ_H with initial position at $\gamma(0) = (p,\xi)$. The projection of $\gamma(t)$ onto M will be called the *null-bicharacteristic* through p in the direction of v. Notice that since γ lies on Σ the tangent vector to this null-bicharacteristic is not only light-like at $t = 0$ but at all points on the interval $-a < t < a$. If $a = +\infty$ we will call this null-bicharacteristic *complete* It is clear from (4.3) that if M is compact the property that a null-bicharacteristic be complete is conformally invariant, even though the parametrization, $t \to \gamma(t)$, depends on the choice of H.

§5.   Suppose now that for all $p \in M$ and all $v \in \partial C_p$ the
null-bicharacteristic through (p,v) is simply periodic.   In
this case we will say that the causal structure defined by the
C_p's has the "Wiederkehr" property (or property "W" for short.)
For instance if a pseudometric on M is Zollfrei the causal
(or conformal) structure defined by it has this property.   We
will show in this section that if M is compact and the causal
structure defined by the C_p's has this property, then the
following assertions are true:

1.   The set of periodic null-bicharacteristics forms a C^∞
     manifold.

2.   This manifold is equipped with an intrinsic contact
     structure.

To see this fix a defining function, H, for Σ, as in §4,
and let $\Sigma_{2\pi}$ be the set of all points, $(p,\xi) \in \Sigma$ with the
property that the integral curve of Ξ_H through (p,ξ) is of
period exactly 2π.   The one-parameter group of diffeomorphisms
of $\Sigma_{2\pi}$ generated by Ξ_H is periodic of period 2π, so it
defines a free action of S^1 on $\Sigma_{2\pi}$.   Since S^1 is compact
this free action is, locally, in the neighborhood of a fixed
orbit, a product action; so the set of orbits forms a Hausdorff
manifold, P, and the point-orbit assignment, π, makes $\Sigma_{2\pi}$
into a principal S^1 bundle over P

(5.1) $$\pi \colon \Sigma_{2\pi} \longrightarrow P.$$

Now let α be the canonical symplectic one-form on the cotangent bundle of M and let $\iota \colon \Sigma_{2\pi} \to T^*M$ be the inclusion map. By the homothety property (4.2)

$$i(\Xi_H)\iota_*\alpha = \iota_*H = 0 \qquad \text{on } \Sigma_{2\pi}$$
$$\text{and}$$
$$D_{\Xi_H}\iota_*\alpha = \iota_*(di(\Xi_H)\alpha + i(\Xi_H)d\alpha) = \iota_*dH = 0 \qquad \text{on } \Sigma_{2\pi}.$$

Thus $\iota_*\alpha$ is "horizontal" with respect to the fibration (5.1) and is S^1 invariant. This means that there exists a one-form, α_0, on P such that

(5.2) $$\iota^*\alpha = \pi^*\alpha_0.$$

It is easy to check directly from (5.2) that α_0 is a contact form on P.

To conclude the proofs of the assertions 1 and 2 above, it is clear that P can be identified with the set of all (unparametrized) periodic null-bicharacteristics. The contact form, α_0, on P depends on the choice of H but it is easy to see from (4.3) that a different choice of H will only change α_0 by multiplying it by a positive function, so the contact structure itself doesn't depend on H.

§6.   Suppose now that one has a family of causal structures on M parametrized by a real parameter, s, where $-\epsilon < s < \epsilon$. We will assume that this family depends smoothly on s and that each member of this family has the "Wiederkehr" property described above. Let Σ_s be the characteristic variety of the s-th member of this family and let H_s be a defining function for it. Without loss of generality we can assume that H_s depends smoothly on s. A main ingredient in our approach to the deformation problem for Zollfrei metrics is the following "symplectic rigidity theorem".

Theorem 6.1.   There exists a smooth function, F_s, on T*M-0, homogeneous of degree zero, and a canonical transformation, Φ_s, from T*M-0 to itself, both depending smoothly on s, such that Φ_0 is the identity transformation, $F_0 = 1$, and

(6.1) $$\Phi_s^* H_s = F_s H_0$$

for all s.

For the proof of this theorem we will need the following lemma:

Lemma 6.2:   Let P be a compact manifold and α_s, $-\epsilon < s < \epsilon$, a family of contact forms on M depending smoothly on s.

Then there exists a smooth function, K_s, and a diffeomorphism, Ψ_s, both depending smoothly on s, such that $K_0 = 1$, $\Psi_0 =$ identity and

$$(6.2) \qquad\qquad \Psi_s^* \alpha_s = K_s \alpha_0 \ .$$

<u>Proof</u> (by the "Moser" trick). Suppose such a Ψ_s and K_s exist. Set $\Xi_s = (\Psi_s^{-1})^* \dot{\Psi}_s$ and $G_s = (\Psi_s^{-1}) \frac{d}{ds} \log K_s$. Then (6.2) is equivalent to

$$(6.3) \qquad\qquad D_{\Xi_s} \alpha_s = -\dot{\alpha}_s + G_s \alpha_s \ .$$

We will assume $i(\Xi_s)\alpha_s = 0$. Then, by Cartan's identity (6.3) reduces to:

$$(6.4) \qquad\qquad i(\Xi_s)d\alpha_s = -\dot{\alpha}_s + G_s \alpha_s$$

which admits a unique solution (Ξ_s, G_s) providing $\alpha_s(\Xi_s) = 0$.

    Returning to the proof of Theorem 6.1 let P_s be the manifold of null-bicharacteristics of H_s and let

$$\pi_s: (\Sigma_s)_{2\pi} \longrightarrow P_s$$

be the fiber mapping, (5.1), associated with H_s. Recall that there exists a canonical contact form, α_s, on P_s such that

(6.5) $$\pi_s^* \alpha_s = \iota_s^* \alpha \,,$$

ι_s being the inclusion map of $(\Sigma_s)_{2\pi}$ into T^*M. Since all the data above depend smoothly on s, the diffeotype of P_s is independent of s. Therefore, by Lemma 6.2, there exists a diffeomorphism

$$\Psi_s \colon P_0 \to P_s$$

and a smooth function, K_s, on P_0 such that $K_0 = 1$, Ψ_0 is the identity and

(6.6) $$\Psi_s^* \alpha_s = K_s \alpha_0.$$

By the covering homotopy property there exists a diffeo-morphism:

$$\tilde{\Psi}_s \colon (\Sigma_0)_{2\pi} \longrightarrow (\Sigma_s)_{2\pi}$$

lifting Ψ_s , i.e. satisfying

(6.7) $$\Psi_s \circ \pi_0 - \pi_s \circ \tilde{\Psi}_s.$$

From (6.6) and (6.7) we get

$$(\tilde{\Psi}_s)^* (\pi_s^* \alpha_s) = (\tilde{K}_s)(\pi_0^* \alpha_0)$$

with $\tilde{K}_S = \pi_0^* K_S$. By (6.5) this is equivalent to

$$(6.8) \qquad\qquad \tilde{\Psi}_S^* \iota_S^* \alpha = \tilde{K}_S \iota_0^* \alpha$$

on $(\Sigma_0)_{2\pi}$. Now extend $\tilde{\Psi}_S$ and $\tilde{K}_S$ to Σ_0 by extending them
homogeneously. Then, (continuing to denote by ι_0 and ι_S
the inclusion maps of Σ_0 and Σ_S into T^*M), it is clear
that (6.8) holds on all of Σ_0.

     Consider next the homothety of Σ_0 onto itself sending
(p, ξ) to $(p, \tilde{K}_S(x, \xi) \xi)$. Call this homothety ρ_S. It is easy
to check that

$$\rho_S^* \iota_0^* \alpha = \tilde{K}_S \iota_0^* \alpha;$$

so we can rewrite (6.8) in the form,

$$\tilde{\psi}_S^* \iota_S^* \alpha = \rho_S^* \iota_0^* \alpha \ ;$$

i.e., setting $\Phi_S = \tilde{\Psi}_S \circ \rho_S^{-1}$, we can rewrite (6.8) in the form,

$$(6.9) \qquad\qquad \Phi_S^* \iota_S^* \alpha = \iota_0^* \alpha.$$

By the co-isotropic imbedding theorem of Weinstein, (see [17],
§42), there is a conic neighborhood, U_S, of Σ_S and a homogeneous canonical transformation

$$\Phi_S : U_0 \longrightarrow U_S,$$

both depending smoothly on s, extending the Φ_S above. Let

$$(6.10) \qquad\qquad \Xi_S = (\Phi_S^{-1})^* \frac{d\Phi_S}{ds}.$$

Because of its homogeneity Ξ_S is a globally Hamiltonian vector field on a conic neighborhood of Σ_0; i.e. it is a Hamiltonian vector field associated with a Hamiltonian function, G_S, defined (and homogeneous of degree one) on a conic neighborhood of Σ_0 and depending smoothly on s. Let us extend G_S to all of T*M-0 and continue to denote this extension by G_S. Let Ξ_S be the Hamiltonian vector field associated with G_S, and notice that what we are now calling Ξ_S is an extension of (6.10) to all of T*M-0. By integrating

$$\Xi_S = (\Phi_S^{-1})^* \frac{d\Phi_S}{ds}$$

with $\Phi_0 = $ identity we get an extension of Φ_S to all of T*M-0. Finally set

$$F_S = \frac{\Phi_S^* H_S}{H_0}.$$

§7. Let M be a compact manifold and let $(d\sigma^2)$ be a Zoll-frei metric on M. By a *Zollfrei deformation* of $(d\sigma)^2$ we will mean a family of Zollfrei metrics, $(d\sigma)^2_s$, $-\epsilon < s < \epsilon$, depending smoothly on the parameter, s, such that $(d\sigma)^2_s = (d\sigma)^2$ when $s = 0$. Given such a deformation, we can expand $(d\sigma)^2_s$ in a Taylor series in s at $s = 0$:

$$(7.1) \qquad (d\sigma)^2_s \sim (d\sigma)^2_0 + \kappa_1 s + \kappa_2 s^2 + \cdots$$

the κ_i's being contravariant symmetric 2-tensors. In this section we will show that rather strong constraints are imposed on the κ_i's by the fact that they define, formally, a Zollfrei deformation of $(d\sigma)^2$. We will show first of all that κ_1 has to satisfy the following condition. For every null-geodesic, γ, of $(d\sigma)^2$,

$$(7.2) \qquad \int_\gamma \kappa_1 \left(\frac{d\gamma}{dt}, \frac{d\gamma}{dt}\right) = 0.$$

Providing this condition is satisfied, the κ_i's can, to a certain extent, be determined recursively from κ_1. What we want to do is reverse this procedure, i.e. given a contra-variant tensor, κ_1, satisfying (7.2), construct a Zollfrei deformation of $(d\sigma)^2$ by solving recursively for the κ_i's. We will see that this involves solving a series of integral

equations similar to (7.2) but with non—zero terms on the right
hand side. This implies that except in dimension three (and
frequently *even* in dimension three) (7.2) is the first of a
countable sequence of infinitesmal obstructions to deform—
ability.

We will begin by deriving (7.2). Let us denote by H_s
the Hamiltonian function associated with the metric $(d\sigma)^2_s$: At
every point, (p,ξ), in the cotangent bundle of M

$$H_s(p,\xi) = \langle\xi,\xi\rangle^*_p.$$

(Here $\langle\cdot,\cdot\rangle^*_p$ is the quadratic form on T^*_p dual to the
Lorentzian quadratic form, $\langle\cdot,\cdot\rangle_p$, on T_p defining the
metric, $(d\sigma)^2_s$, at p.) The function, H_0, will be simply
denoted by H; i.e. H will be the Hamiltonian of the original
metric, $(d\sigma)^2$. By Theorem 6.1 there exists a canonical
transformation, Ψ_s, of T^*M-0 onto itself and a smooth
homogeneous function of degree zero, F_s, on T^*M-0 such that

(7.3) $$\Psi^*_s H_s = F_s H.$$

In addition we can assume that F_s and Ψ_s depend smoothly on
s and that Ψ_0 is equal to the identity and F_0 equal to
one. Let Ξ be the Hamiltonian vector field, $\dfrac{d\psi_s}{ds}$, at $s = 0$.
This vector field is globally Hamiltonian. In fact we can

express it in terms of the canonical one form, α, on T^*M as
the Hamiltonian vector field associated with the function
$G = -\iota(\Xi)\alpha$. Differentiating (7.3) and setting $s = 0$ we
obtain:

$$(7.4) \qquad\qquad \dot{H} + \{G,H\} = \dot{F}H.$$

Now let Ξ_σ be the Hamiltonian vector fields corresponding to
H, and let Σ be the subset of the cotangent bundle where
$H = 0$. Recall that the null-geodesics of $(d\sigma)^2$ are the projections onto M of the integral curves of Ξ_σ lying on Σ.
On Σ, (7.4) reduces to:

$$(7.5) \qquad\qquad \dot{H} = D_{\Xi_\sigma} G.$$

Integrating this expression over a null-bicharacteristic, γ,
the right hand side is zero and we get

$$(7.6) \qquad\qquad \int \dot{H}(\gamma(t))dt = 0.$$

It is easy to see that this is equivalent to (7.2).

Let $\Sigma_{2\pi}$ be the set of points, (p,ξ), in the cotangent
bundle of M such that the null-bicharacteristic of H
through (p,ξ) is of period 2π. Let P be the manifold of
(unparametrized) null-geodesics of $(d\sigma)^2$. By (5.1) $\Sigma_{2\pi}$

fibers over P, each fiber being a periodic null-bicharacter-
istic. Moreover, the fibration, $\pi: \Sigma_{2\pi} \longrightarrow P$, is a principal
S^1-fibration, the action of S^1 being given by

$$S^1 \ni e^{it} \longrightarrow \exp t\Xi_\sigma .$$

(This action is well-defined since $\exp t\Xi_\sigma$ is the identity
when $2\pi = t$.)

Suppose now that we are given a function, $\dot{H}$, on T^*M
which is quadratic on each cotangent fiber and satisfies (7.6).
Consider the equation (7.5). A putative solution of this
equation is given by the variation of constants formula

$$(7.7) \qquad G = -\frac{1}{2\pi} \int_0^{2\pi} \left[\int_0^t (\exp s\Xi_\sigma)^* \dot{H} ds \right] dt.$$

(It is easy to check that this is a solution of (7.5) if and
only if $\dot{H}$ satisfies (7.6).) The formula, (7.7), only defines
G on $\Sigma_{2\pi}$; however, G can be extended to all of Σ by
requiring it to be homogeneous of degree one. Finally if one
extends G arbitrarily to be a homogeneous function of degree
one on all of T^*M-0, and sets

$$(7.8) \qquad\qquad \dot{F} = \frac{\{G,H\}+\dot{H}}{H}$$

then (7.4) will be identically satisfied on all of T^*M-0.

Let us now examine the problem of extending the infin-
itesmal deformation, $\dot{H}$, (or κ_1) of $(d\sigma)^2$ to the "n-th formal
neighborhood" of 0 in the parameter space, $-\epsilon < s < \epsilon$. By such
an extension we will mean the following three pieces of data:

1) A canonical transformation, Ψ_s, of T*M-0 onto itself.
2) A smooth homogeneous function of degree zero, F_s, on
   T*M-0.
3) A smooth function, H_s, globally defined on T*M and
   quadratic on each cotangent fiber.

These data are required to depend smoothly on s, take on the
appropriate initial values (i.e. $F_0 = 1$, $H_0 = H$, $(dH_s/ds)_0 = \dot{H}$
and Ψ_0 = identity) and, most important of all, to satisfy the
n-th order deformation equation:

$$(7.9) \qquad\qquad \psi_s^* H_s = F_s H_0 \text{ modulo } 0(s^{n+1}).$$

For instance, for n = 1, (7.9) is equivalent to (7.4) if we set:

$$(7.10) \qquad \psi_s = \exp s \, \Xi_G \,, \qquad F_s = 1 + s\dot{F} \,, \qquad H_s = H + s\dot{H} \,.$$

Suppose now that we have succeeded in extending $\dot{H}$ to the
(n-1)st formal neighborhood of s = 0 in deformation space,
i.e. suppose we have produced a canonical transformations, Ψ_s,

and functions F_s and H_s with the properties described above such that (7.9) is satisfied modulo $O(s^n)$. Then we can write

(7.11) $$(\Psi_s)^* H_s = F_s H + s^n R_n + O(s^{n+1}).$$

Let us try to find smooth functions, G_n, F_n and H_n on T^*M-0, such that, on each cotangent fiber, G_n is homogeneous of degree one, F_n homogeneous of degree zero and H_n a homogeneous quadratic polynomial and such that (7.9) is satisfied with Ψ, F and H replaced by

(7.12) $$\Psi' = \Psi \circ (\exp s^n \Xi_{G_n}),$$
$$F' = F + s^n F_n,$$
$$H' = H + s^n H_n.$$

It is easy to see that, because of (7.11), (7.9) reduces to

(7.13) $$\{G_n, H\} = F_n H + H_n + R_n.$$

Let γ be a null-bicharacteristic of H of period 2π. Then integrating (7.13) over γ we get for R_n the n-th order integrability conditions

$$(7.14) \qquad \int_0^{2\pi} H_n(\gamma(t)dt = - \int_0^{2\pi} R_n(\gamma(t))dt.$$

Conversely suppose that we can find an H_n such that (7.14) holds for all γ. Then we can define G_n as before (see (7.7)) by the "variation of constants formula" on $\Sigma_{2\pi}$:

$$(7.15) \qquad G_n = - \frac{1}{2\pi} \int_0^{2\pi} dt \int_0^t (\exp s\Xi_\sigma)^*(H_n + R_n)ds$$

and then, in the same way as before, extend it to all of T*M-0. Finally letting

$$(7.16) \qquad F_n = \frac{\{G_n, H\} - (H_n + R_n)}{H}$$

we get (7.13) to hold on all of T*M-0.

The solvability of (7.14) can be formulated in terms of an integral operator which we will study in more detail in part 4. Let $S^2(T)$ be the symmetric tensor product of the tangent bundle of M with itself and let $C^\infty(S^2(T))$ be the space of smooth global sections of $S^2(T)$. An element of $C^\infty(S^2(T))$ can be regarded as a function on T*M which is quadratic on each cotangent fiber. By restricting this function to Σ_{2n}, we get a C^∞ function on Σ_{2n}. This procedure defines for us a map

$$(7.17) \qquad C^\infty(S^2(T)) \to C^\infty(\Sigma_{2\pi}).$$

On the other hand, associated with the principal fibration, $\pi: \Sigma_{2\pi} \longrightarrow P$, (see (5.1)), there is an operation of fiber integration

$$(7.18) \qquad\qquad C^{\infty}(\Sigma_{2\pi}) \longrightarrow C^{\infty}(P),$$

and, composing (7.17) and (7.18), we get a "generalized x-ray transform"

$$(7.19) \qquad\qquad R_{\sigma}: C^{\infty}(S^2(T)) \longrightarrow C^{\infty}(P).$$

Coming back to (7.14), as one varies $\gamma \in P$ the right hand side of (7.14) varies smoothly, so we can regard the right hand side of (7.14) as defining a smooth function, $(r_n)_{av} \in C^{\infty}(P)$. Solving (7.14) for all γ amounts to finding a smooth section, h_n, of $S^2(T)$ such that

$$(7.20) \qquad\qquad R_{\sigma}h_n = (r_n)_{av} .$$

For instance we will prove in §10 that if one takes $(d\sigma)^2$ to be the Einstein metric on $M_{2,1}$ or its double cover, then (7.19) is surjective. Therefore, we have proved

<u>Theorem 7.1</u>. Let M be $M_{2,1}$ (or its double cover) and let $(d\sigma)^2$ be the Einstein metric. Then for every κ_1 satisfying (7.2) there exists a formal power series solution

$$(7.21) \qquad\qquad (d\sigma)^2_s = (d\sigma)^2 + s\kappa_1 + s^2\kappa_2 + \cdots$$

to the deformation problem.

§8.   Unfortunately the formal power series (7.21) doesn't seem to converge except for special choices of κ_1. In this section we will discuss a modified form of the iteration scheme, (7.13)-(7.16), (based on "Newton's method"). In this scheme the error at stage n will be of order s^k, $k = n^2$, rather than s^n. Moreover, by "filtering out high frequencies" in the successive correction terms a la Nash-Moser, one can make this scheme converge to a limit which is C^∞ both in the manifold variables and in the deformation parameter s. Unfortunately this scheme requires a lot more technical machinery than the simple linear iteration scheme: (7.13)—(7.16). In (7.13)—(7.16) one had to solve at each stage the equation

$$(7.20): \qquad\qquad R_\sigma h_n = (r_n)_{av} .$$

Therefore one can implement this scheme by showing that R_σ is surjective. In the scheme described below the operator playing the role of R_σ will change from stage to stage and at each stage will depend on the deformation parameter s. Therefore to implement this scheme one has to prove the simultaneous surjectivity of all these operators. In fact if one wants to couple this scheme with Nash-Moser one has to show that these operators satisfy *uniform* a priori estimates of a type which we will describe below.

Suppose as in §7 that we are given a function, $\dot{H}$, on the cotangent bundle of M which is a quadratic polynomial on each cotangent fiber and satisfies (7.6). We will try to extend this infinitesmal deformation to the 2^n-th formal neighborhood of $s = 0$ in deformation space in n successive stages. At the n-th stage we will require the following three pieces of data.

1. A canonical transformation, Ψ_s, of T*M-0 onto itself.
2. A smooth homogeneous function of degree zero, F_s, on T*M-0.
3. A smooth function, H_s, on T*M-0 whose restriction to each cotangent fiber is a quadratic polynomial.

As before we will require that these data depend smoothly on s, satisfy appropriate initial conditions (Ψ_0 = identity, $F_0 = 1$, $H_0 = H$, $(dH_s/ds)_{s=0} = \dot{H}$) and satisfy the n-th order deformation equation

$$(8.1) \qquad \Psi_s^* H = F_s H_s + O(s^k) , \qquad \text{with } k = 2^n.$$

(Notice that there is one other difference between (8.1) and (7.9) besides the "2^n" in the error term: the positions of H and H_s have been interchanged.)

Suppose that we have constructed data, Ψ_s, F_s and H_s, satisfying (8.1). We want to replace this with data, Ψ_s', F_s'

and H'_S satisfying (8.1) with $n+1$ in place of n. Moreover, we want a "fast" procedure for manufacturing the primed data from the unprimed data. (In other words we don't want just to apply the procedure of §7 2^n times.) As in §7 we will make the Ansätze:

$$(8.2) \qquad \begin{aligned} \Psi'_S &= \Psi_S \circ \exp s^k \Xi_{G_n} \\ F'_S &= F_S + s^k F_n \\ H'_S &= H_S + s^k H_n \end{aligned}$$

but we will now allow G_n, F_n and H_n to be function of the deformation parameter, s, as well as the cotangent variables. Setting

$$(8.3) \qquad R_n = \frac{\Psi_S^* H - F_S H_S}{s^k}$$

and writing F for F_S and Ψ for Ψ_S the $(n+1)$-st deformation equation becomes

$$(8.4) \qquad \{G_n, \Psi^* H\} = (F_n/F)\Psi^* H + FH_n + R_n + 0(s^k).$$

We will attempt to solve this equation *exactly*, i.e. we will look for F_n, G_n and H_n depending smoothly on the (surpressed) parameter s such that

(8.5) $\{G_n, \Psi^*H\} = (F_n/F)\Psi^*H + FH_n + R_n.$

Before we start lets point out a few features of the function

(8.6) $H_\Psi = \Psi^*H .$

This function no longer has the property that its restriction
to each cotangent fiber is a quadratic polynomial. However,
for small s and for every $p \in M$ the set

$$(\Sigma_\Psi)_p = \{\xi \in T_p^*, \ H_\Psi(p,\xi) = 0\}$$

is a convex cone in T_p^*-0. If we assign to each point $p \in M$
the dual cone in T_p-0, we get a causal structure on M in the
sense of §4. Moreover, since Ψ maps null-bicharacteristics of
H onto null-bicharacteristics of H_Ψ, this causal structure
has the Wiederkehr property described in §5, i.e., the null-
bicharacteristics of H_Ψ are simply periodic and form a smooth
manifold, P, diffeomorphic to the manifold of null-geodesics
of $(d\sigma)^2$.

    Returning to the equation (8.5) let Ξ_Ψ be the
Hamiltonian vector field corresponding to the Hamiltonian
function H_Ψ. To solve (8.5) it is sufficient to solve

(8.7) $\{G_n, H_\Psi\} = F H_n + R_n$

on $(\Sigma_\Psi)_{2\pi}$. As in (7.15) we would like to solve (8.7) by variation of constants, i.e., by setting

$$(8.8) \qquad G_n = \frac{1}{2\pi} \int_0^{2\pi} dt \int_0^t (\exp s\Xi_\Psi)^*(FH_n + R_n)ds.$$

However, (8.8) implies (8.7) if and only if

$$(8.9) \qquad \int_0^{2\pi} (\exp t\Xi_\Psi)^* FH_n \; dt = - \int_0^{2\pi} (\exp t\Xi_\Psi)^* R_n \; dt$$

on $(\Sigma_\Psi)_{2\pi}$. (Notice by the way that (8.9) involves the surpressed deformation parameter s. It is really a family of equations, one for each s on the interval $-\epsilon < s < \epsilon$.)

As in §7 we can interpret (8.9) as an integral equation. As we let H_n vary, the left hand side of (8.9) defines an integral operator

$$R_{\Psi,F}: \; C^\infty(S^2(T)) \longrightarrow C^\infty(P).$$

The right hand side of (8.9) can be viewed as a smooth function, $(r_n)_{av}$, on P; and, if we let h_n be the symmetric co-variant two-tensor on M corresponding to the function H_n, (8.9) can be written in the more appealing form,

$$(8.10) \qquad R_{F,\Psi} h_n = (r_n)_{av} ,$$

(which one can view as the "twisted" analogue of (7.20).)

    Lets briefly summarize what we've done so far and draw some conclusions from it:

1.   Let Ψ be a canonical transformation of T*M-0 onto itself and let F be a smooth homogeneous function of degree zero on T*M-0. With this data is associated an integral operator

$$(8.11) \qquad\qquad R_{\Psi,F} \colon C^\infty(S^2(T)) \longrightarrow C^\infty(P).$$

2.   For the Newton's method outlined in this section to work it is neccessary that (8.11) be surjective for Ψ sufficiently close to the identity and F sufficiently close to one.

    In fact Nash-Moser requires a little more; it requires Sobolev estimates of the form

$$(8.12) \qquad \|R_{\Psi,F}^r h\|_{s+m} + \|h\|_t \geq C(s,t)\|h\|_s, \qquad s > t,$$

uniform in (Ψ,F) for Ψ close to the identity and F close to one. (A proof of such estimates will be given in §17. See (17.40).)

PART III

ZOLLFREI DEFORMATIONS OF $M_{2,1}$

§9. In the third part of this paper we will investigate the
implication of the results of part two for deformations of the
standard Einstein metric on $M_{2,1}$ and its double cover. There
is one feature of $M_{2,1}$ which will enormously simplify this
investigation, namely the fact that it possesses a very large
group of conformal symmetries (the ten dimensional group,
$Sp(2,\mathbb{R})$). To exploit this fact we will show that the x-ray
transform for $M_{2,1}$ has what the physicists would call a
"manifestly $Sp(2,\mathbb{R})$-invariant" description. Before we do so,
however, we will pause for a moment to give (for the first time
in this monograph) a careful definition of a term which we used
extensively in part two, the term "infinitesmal conformal
deformation": Let M be an n-dimensional manifold and g a
pseudo-Riemannian metric on M. Associated with g is its
volume form, $vol(g)$; and under the conformal change, $g \to \rho g$,
it transforms into the form, $\rho^{n/2} vol(g)$. Therefore, if we let
λ_M be the n-th root of the volume bundle, we can identify λ_M^2
with the line bundle of metrics on M which are conformally
equivalent to g. In particular we have an imbedding

$$\lambda_M^2 \rightarrow S^2(T^*)$$

and a splitting

$$S^2(T^*) = S^2(T^*)_0 \oplus \lambda_M^2 ,$$

$S^2(T^*)_0$ being the bundle of *traceless* symmetric contravariant tensors of degree 2. Now let g_t be a deformation of g depending smoothly on the parameter, t, with $g_0 = g$. Let $\dot g_0$ be the $S^2(T^*)_0$ component of the tensor

$$(\frac{d}{dt} g_t) \qquad (t = 0) .$$

Notice that if we replace g_t by the conformally equivalent deformation, $\rho_t g_t$, $\dot g_0$ gets changed to $\rho_0 \dot g_0$. Therefore, we see that a conformal deformation of M gives rise to a morphism of vector bundles

$$\lambda_M^2 \rightarrow S^2(T^*)_0 ,$$

or, alternatively, a section of

(9.1) $$S^2(T^*)_0 \otimes \lambda_M^{-2} .$$

Thus we are led to define the *infinitesmal conformal defor-mations* of (M,g) *to be the sections of* (9.1). Notice that if v is a vector field, the traceless part of $D_v(\rho g)$ is equal to ρ times the traceless part of $D_v g$; so v defines a morphism of vector bundles, $\lambda_M^2 \longrightarrow S^2(T^*)_0$, and hence a section, $\kappa(v)$, of (9.1). As we vary v we get a linear mapping

$$(9.2) \qquad \kappa: \Gamma(T^*) \longrightarrow \Gamma(S^2(T^*) \otimes \lambda_M^{-2}) \ .$$

The image of κ is the space of *trivial* infinitesmal conformal deformations of g. (Thus, strictly speaking, when one speaks of infinitesmal conformal deformations of g, one doesn't mean sections of (9.1). One means the *projections* of these sections onto the cokernel of κ.)

Let's now return to $M = M_{2,1}$. From now on we will work almost exclusively with the Kleinian description of $M_{2,1}$ outlined in §2. Namely we will think of $M_{2,1}$ as the Grassmannian of Lagrangian subspaces of $\mathbb{R}^4$. To avoid the usual confusion in nomenclature that arises when dealing with Grassmannians we will denote the points of $M_{2,1}$ by lower case Roman letters like p, q, etc. and the Lagrangian spaces which they represent by V_p, V_q, etc. Let V be the tautology bundle whose fiber at p is V_p. Then the tangent bundle of $M_{2,1}$ is $T = S^2(V^*)$; and the conformal structure on M is defined by defining the forward light cone in T_p to be the set

$$(9.3) \qquad \{\mu \otimes \mu, \ \mu \in V_p^*.\}$$

Notice that if ω is a non-zero element of $\Lambda^2[V]^*$ one can define an inner product, g_p, on T_p by defining, for $S \in T_p = S^2(V^*)$

$$\omega(J_S v, w) = S(v,w), \; \forall \; v,w \in V$$

and

$$\det(J_S) = g_p(S,S).$$

If one replaces ω by $\lambda\omega$ in this definition g_p gets changed to $\lambda^{-2}g_p$; so there is a canonical identification

$$(9.4) \qquad\qquad \lambda_M \cong \Lambda^2[V].$$

Since the tangent bundle of $M_{2,1}$ is $S^2(V^*)$, the cotangent bundle is $S^2(V)$, and the bundle of traceless symmetric contravariant two-tensors is $S^4(V)$. Thus the space of infinitesmal conformal deformations of $M_{2,1}$ is the space of sections of the bundle

$$(9.5) \qquad\qquad S^4(V) \otimes (\Lambda^2[V^*])^2.$$

Next lets see what the x-ray transform looks like in this setting: We showed in §2 that the null-geodesics of $M_{2,1}$ are in one-one correspondence with the points of $\mathbb{RP}^3$ or, alternatively, with the one-dimensional subspaces of $\mathbb{R}^4$. In the

Kleinian picture one sees this correspondence by noticing that the null-geodesics are just the one-dimensional "Shubert cycles" of the Lagrangian Grassmannian. More precisely given a one-dimensional subspace, W, of $\mathbb{R}^4$ the Shubert cycle associated with W

$$(9.6) \qquad \{p \in M_{2,1} \; , \; W \subset V_p\},$$

is an algebraically imbedded $\mathbb{RP}^1$. Indeed, if V_p contains W, then V_p is contained in the symplectic orthocomplement, $W^\perp$, of W; so the set, (9.6), is isomorphic to the one-dimensional projective space:

$$(9.7) \qquad \text{Proj}(W^\perp/W).$$

It is easy to check that the sets (9.6) are exactly the null-geodesics of compactified Minkowski space.

Now let f be a section of (9.5) and let μ be a non-zero element of W. Consider a point p in the set (9.6); i.e., a point p in $M_{2,1}$ with $W \subset V_p$. The projection mapping, $V_p \to V_p/W$, gives rise to a mapping $S^4(V_p) \to S^4(V_p/W) = (V_p/W)^4$ and by contraction by $\mu \otimes \mu$ one gets a map

$$(9.9) \qquad (\Lambda^2[V_p]^*)^2 \longrightarrow (\Lambda^1[V_p/W]^*)^2 = (V_p/W)^{-2}.$$

Combining these two maps one can convert $f(p)$ (which is an element of $S^4(V_p) \otimes (\Lambda^2[V_p]^*)^2)$ into $\tilde{f}_\mu(p)$, an element of the one-dimensional space $(V_p/W)^2$. If we globalize this construction by varying p in the set (9.6) we obtain a homogeneous function of degree -2, $\tilde{f}_\mu$, on the two-dimensional vector space (with origin deleted) $(W^\perp/W)-\{0\}$. Notice, however, that $W^\perp/W$ comes equipped with a symplectic form, ω_W; and since we are in two dimensions, ω_W can be thought of as a volume form. Therefore we can form the *residue* of $\tilde{f}_\mu$ with respect to this volume form to obtain a numerical quantity:

$$(9.10) \qquad\qquad \mathrm{Res}(\tilde{f}_\mu \omega_W) = R(f,\mu)_W.$$

(See the appendix at the end of this section.) It is clear from (9.9) that this quantity depends quadratically on μ.

Now let L be the canonical line bundle of $\mathbb{RP}^3$ and q the point in $\mathbb{RP}^3$ corresponding to the one-dimensional subspace, W, of $\mathbb{R}^4$. The fiber of L at q is W; so the quadratic form

$$\mu \longrightarrow R(f,\mu)_W, \qquad \mu \in W,$$

is an element of $(L^*)_q^2$. Globalizing this construction by letting q vary in $\mathbb{RP}^3$, we obtain from f a global section,

Rf, of $(L^*)^2$. Summarizing we have exhibited the existence of a transform

$$(9.11) \qquad R: \Gamma(S^4(V) \otimes (\Lambda^2[V^*])^2) \longrightarrow \Gamma(L^*)^2$$

which integrates data on the right hand side over the null-geodesics of $M_{2,1}$. This is, of course, just the x-ray transform, (7.19), in disguise; the description we have just given of it displays clearly its $Sp(2,\mathbb{R})$-invariant character.

In part one we showed that an infinitesmal conformal deformation, f, of $M_{2,1}$ which corresponds to a deformation of "Zollfrei" type has to satisfy the integrability condition: Rf = 0. We also proved the converse (modulo some questions in analysis which we've deferred to part four.) It is clear from the results in part one that the trivial deformations (the image of κ in (9.2)) satisfy this condition. However, what else satisfies this condition? We will give a definitive answer to this question in the next section, but, for the moment, we will show that there are other elements in the kernel of R besides those in the image of κ.

In keeping with the notation which we used in §2 we will denote by $\mathbb{C}^4$ the complexification of $\mathbb{R}^4$, by $\omega_{\mathbb{C}}$ the $\mathbb{C}$-linear extension of ω to $\mathbb{C}^4$, by $(M_{2,1})_{\mathbb{C}}$ the space of complex Lagrangian subspaces of $\mathbb{C}^4$ and by $M_{2,1}^+$ the space of positive-definite Lagrangian subspaces of $\mathbb{C}^4$. Notice that the

closure of $M_{2,1}^+$ is the space of positive *semi-definite*
Lagrangian subspaces of $\mathbb{C}^4$ and contains $M_{2,1}$ as one boundary component. The vector bundles, V, $S^2(V)$, $\Lambda^2[V]$ etc.
have holomorphic extensions to $(M_{2,1})_\mathbb{C}$, and we will continue
to use our old notation for them.

Suppose now that f is a section of (9.5) which is
defined and holomorphic on an open subset of $(M_{2,1})_\mathbb{C}$ which
contains the closure of $M_{2,1}^+$.

<u>Theorem 9.1</u>. The restriction, f_0, of f to $M_{2,1}$ is in the
kernel of R.

<u>Proof</u>: Fix a one dimensional subspace, W, of $\mathbb{R}^4$ and con-
sider the *complex* Shubert cycle

$$(9.12) \qquad\qquad \{p \in (M_{2,1})_\mathbb{C}, \ W \subset V_p\}.$$

This is a complex projective line containing the real projective line (9.6). Moreover, (9.6) divides (9.12) into two hemi-
spheres, the upper hemisphere being the intersection of (9.12)
with $M_{2,1}^+$. As we will see in the appendix, the residue,
(9.10), can be interpreted as the integral of a certain one-
form over the contour (9.6). However, if f_0 extends holo-
morphically over the upper hemisphere, the same will be true of

this one-form; and so by the Cauchy integral formula its integral over the boundary will be zero.   Q.E.D.

This result can be strengthened considerably. If f is a holomorphic section of (9.5) over $M_{2,1}^+$, it has a hyperfunction trace, f_0, on $M_{2,1}$. One can show that if this hyperfunction trace is smooth, then it is in the kernel of R.

APPENDIX (the residue construction). Let V be a vector space of even dimension, n, with a fixed volume form, ω, and let f be a function on V—0 which is homogeneous of degree -n. We will denote by Ξ the vector field on V associated with the one-parameter group of homotheties

$$(9.13) \qquad\qquad t \longrightarrow e^t \ \text{Identity}.$$

Consider the (n-1)-form

$$\mu_f = \iota(\Xi) f \omega$$

on V-0. It is clearly invariant with respect to the action, (9.13), of $\mathbb{R}$. Moreover, its interior product with the vector field, Ξ, is zero since

$$\iota(\Xi)\mu_f = \iota(\Xi) \ \iota(\Xi) f \omega = 0.$$

Therefore, it is *basic* with respect to the fibration

$$V\text{-}0 \xrightarrow{\ \pi\ } \text{Proj}(V);$$

that is, there exists an (n-1)-form, ν_f, on Proj(V) such that $\pi^*\nu_f = \mu_f$. One defines the residue, Res(f,ω) to be the integral of μ_f over Proj(V). (For a more detailed account, see [12]).

§10.   In this section we will show how to deduce almost all
the results which will be needed about the transform, R, from
the fact that R is an $Sp(2,\mathbb{R})$ invariant object. Unfortunately, in the course of the next few paragraphs we will need
to cite a number of results in representation theory which are
very technical (even to state). An extremely good reference
for this material is Knapp: *Representation theory of semi-
simple Lie groups, an overview based on examples* (Princeton U.
Press, 1986).

      Let G be a connected semisimple Lie group and P a
parabolic subgroup of G. Given an irreducible representation,
χ, of P on a finite dimensional vector space, we will denote
by E_χ the vector bundle over the coset space, M = G/P, induced from χ, and by $\mathrm{Ind}_G\chi$ the representation of G on
sections of this vector bundle. Let g be the Lie algebra of
G, $U(g)$ the universal enveloping algebra, and K a maximal
compact subgroup of G. By definition a smooth section of E_χ
is K-finite if there exists a finite-dimensional K-invariant
subspace of $\Gamma(E_\chi)$ containing it. Let $\Gamma(E_\chi)_0$ be the space
of all K-finite sections of E_χ. This space is a finitely-
generated K-$U(g)$ module, and is dense in $\Gamma(E_\chi)$ with respect
to the C^∞ topology.

      The first fact that we will need about these objects is
the following.

<u>Theorem</u>. $\Gamma(E_\chi)_0$ is an "Artinian" object in the category of finitely-generated $K-U(g)$ modules: i.e., there exists a *maximal* chain

$$\{0\} \subset M_1 \subset M_1 \subset \cdots \subset M_r \subset \Gamma(E_\chi)_0$$

of $K-U(g)$ submodules. This chain is not unique, but the irreducible quotients, $M_i|M_{i-1}$, i = 1,...,r+1 (and the multiplicities with which they occur) *are* unique.

<u>Proof</u>:   See Knapp, page 373, Corollary 10.39.

Thanks to some (very deep) recent developments in the theory of Verma modules, one can considerably strengthen this result: one can, in principle, write down formulas for the multiplicities of these irreducible quotients (see [44]). The explicit details have only been carried out for a few groups. Fortunately, however, one of the groups for which this has been done is $SP(2,\mathbb{R})$ (loc. cit., pages 253–255). In fact, in some unpublished work, Luis Casian has pushed these computations one step further. For $Sp(2,\mathbb{R})$ he has actually computed the lattice structure of the lattice of $K-U(g)$ submodules of $\Gamma(E_\chi)_0$. It turns out that even though an infinite number of representations are involved, a very small number of actual lattice configurations occur.[*] We will, henceforth, refer to these as the *Casian diagrams* of the induced representations of $Sp(2,\mathbb{R})$. We will now describe the Casian diagrams of the representations associated with R:

Let V_0 be a fixed Lagrangian subspace of $\mathbb{R}^4$ and let W_0 be a fixed one-dimensional subspace of V_0. The two maximal parabolics, P_1 and P_2, of $Sp(2,\mathbb{R})$ are the stabilizers of V_0 and W_0 and the coset spaces associated with them are $M_{2,1}$ and $\mathbb{R}P^3$. The domain of the integral transform, R, is, by (9.5), the space of sections of the induced bundle, E_{χ_1}, associated with the standard representation, χ_1, of P_1 on the space

$$S^4(V_0) \otimes (\Lambda^2[V_0]^*)^2.$$

The image of R is, by (9.9), contained in the space of sections of the induced bundle, E_{χ_2}, associated with the standard representation, χ_2, of P_2 on the space, $(W_0^*)^2$. The Casian diagrams for these representations are:

---

[*] Casian's results are not yet published; however, results similar to his (though not quite as definitive) can be found in [28] and [44].

1.   For χ_1 the diagram

Figure 1

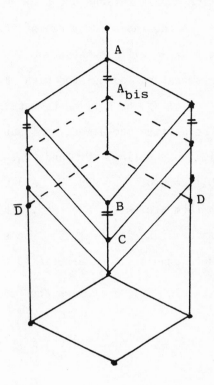

The slashed arrow indicates that the module represented by the lower dot is of finite codimension in the module represented by the upper dot.)

2.   For χ_2 the diagram

Figure 2

(i.e., the module associated with χ_2 contains one finite dimensional irreducible submodule and its quotient is also irreducible.)

Let us now try to fit these diagrams to the geometric facts at our disposal. We will begin with the second (and simplest) of these diagrams: The representation $\text{Ind}_G\chi_2$ is the representation of $Sp(2,\mathbb{R})$ on the space of homogeneous functions of degree 2 from $\mathbb{R}^4$-0 to $\mathbb{C}$. This space contains the homogeneous polynomials of degree 2 as an irreducible subspace, and, hence, by the diagram above, the quotient space is also irreducible. (Notice, by the way, that the finite dimensional irreducible occurring here is just the *adjoint* representation of $Sp(2,\mathbb{R})$.)

From the diagram we can infer

Theorem 10.1. The transform, R, is surjective at the "k-finite level".

Indeed it's clear that its range contains more than just the quadratic polynomials; so its range has to be everything. Coupled with the Sobolev estimates of part four, this theorem implies *that* R *is surjective at the* C^∞ *level*, (and, in fact, surjective on appropriate Sobolev-space completions of $\Gamma(E_{\chi_1})$ and $\Gamma(E_{\chi_2})$.)

§11.   The geometric content of the first diagram is unfortunately not quite so transparent. From the top row of the
diagram one can read off the fact that the representation
$\text{ind}_G \chi_1$ contains a *maximal* subrepresentation. It is clear from
Theorem 10.1 what this representation is, viz. the preimage
with respect to R of the maximal subrepresentation of
$\text{ind}_G \chi_2$. However, the structure of the diagram below this level
can't be explained as easily. In fact some rather subtle facts
about three-dimensional conformal geometry are needed to under-
stand the rest of the diagram. To explain these facts, let
(M,g), for the moment, be an arbitrary three-dimensional
pseudo-Riemannian manifold. We recall that the basic conformal
invariant associated with g, the Weyl conformal curvature
tensor, vanishes identically in dimension three. In its place,
however, there is another conformal invariant, the *Cotton
tensor*

$$(11.1) \qquad C_{ijk} = R_{ij,k} - R_{ik,j} + \tfrac{1}{4}(g_{ik}R_{ij} - g_{ij}R_{ik})$$

which plays a role in three dimensional conformal geometry
analogous to the Weyl conformal curvature tensor in dimensions
greater than three. For example $C_{ijk} \equiv 0$ if and only if g
is conformally flat. As a global object the Cotton tensor is a
section of $T^* \otimes \Lambda^2(T^*)$. However, in three dimensions, the

Hodge star operator gives one a conformally invariant identification of $\Lambda^2(T^*)$ with $T^* \otimes \lambda_M$; and so one can think of
(11.1) as a section of $T^* \otimes T^* \otimes \lambda_M$. In fact it turns out to
be symmetric and traceless, viewed as a section of this bundle;
so it is actually a section of

$$S^2(T^*)_0 \otimes \lambda_M \; ,$$

or of the conformally isomorphic bundle:

(11.2) $$S^2(T)_0 \otimes \lambda_M^5.$$

Notice that sections of this bundle pair with sections of the
bundle of infinitesmal conformal deformations, (9.1), to give
sections of λ_M^3, i.e., *volume forms*. Thus if M is oriented
(which is *not* the case for $M_{2,1}$) one can define a linear
functional on the space of infinitesmal conformal deformations
of M by pairing with the Cotton tensor to get a bundle
morphism

$$S^2(T^*)_0 \otimes \lambda_M^{-2} \longrightarrow \lambda_M^3,$$

and then integrating over M. This functional turns out to be
the gradient of the Chern-Simons invariant (See [5].) We call
attention to this fact because, as we will see in part five, a

recipe like this computes the gradient of the "Floquet determinant" along a conformal deformation of $M_{2,1}$.

The Cotton tensor satisfies a Bianchi identity similar to
the usual Bianchi identity for the curvature tensor. This
identity says that a certain expression in the first covariant
derivatives of the Cotton tensor vanishes identically. This
expression is a contravariant tensor of degree one and conformal weight four (i.e., there exists a (conformally invariant)
first order differential operator

$$(11.3) \qquad \Gamma(S^2(T^*)_0 \otimes \lambda_M) \longrightarrow \Gamma(T^* \otimes \lambda_M^3)$$

which is *zero* when applied to the Cotton tensor.) To summarize, three dimensional conformal geometry gives rise to the
sequence

$$(11.4) \qquad CM \longrightarrow \Gamma(S^2(T^*)_0 \otimes \lambda_M) \longrightarrow \Gamma(T^* \otimes \lambda_M^3)$$

"where CM is the space of conformal structures on M."
Consider now the first variation of this sequence with respect
to a fixed conformal structure, (M,g). This is a sequence of
linear mappings

$$\Gamma(S^2(T^*)_0 \otimes \lambda_M^{-2}) \xrightarrow{\sigma} \Gamma(S^2(T^*)_0 \otimes \lambda_M) \xrightarrow{\beta} \Gamma(T^* \otimes \lambda_M^3) \ .$$

Combining with (9.2) we obtain the sequence

$$(11.5) \qquad 0 \longrightarrow \Gamma(T) \xrightarrow{\kappa} \Gamma(S^2(T^*)_0 \otimes \lambda_M^{-2})$$
$$\xrightarrow{\sigma} \Gamma(S^2(T^*)_0 \otimes \lambda_M) \xrightarrow{\beta} \Gamma(T^* \otimes \lambda_M^3) \longrightarrow 0.$$

Notice that κ and β are first order differential operators, and that σ is a *third* order differential operator by (11.1). Moreover κ, σ and β are conformally invariant, since the whole set-up above is conformally invariant. As we pointed out above, $\beta \circ \sigma = 0$. In general $\sigma \circ \kappa \neq 0$; however there is one important situation for which $\sigma \circ \kappa = 0$. If (M, g) is conformally flat, then its Cotton tensor vanishes. Moreover, if we deform (M, g) by deforming M by an isotopy, the deformed space, (M, g_t), is conformally flat; so *its* Cotton tensor vanishes. Therefore, by differentiating by t and setting t = 0, we see that $\sigma \circ \kappa = 0$. In other words we have proved:

<u>Proposition 11.1.</u> If (M, g) is conformally flat (11.5) is a *complex*.

This complex is studied in detail by Gasqui and Gold-schmidt in [10]. Among other things they show that, at the sheaf level, (11.5) is exact. Hence they are able to conclude:

Theorem 11.2. (Gasqui-Goldschmidt) The complex, (11.5),
computes the cohomology of M with values in the sheaf of
conformal vector fields. In particular the first cohomology
group of (11.5) is the space of infinitesmal conformally flat
deformations of (M, g).

Coming back to $M_{2,1}$, we can, by means of vector bundle
isomorphism discussed in section nine, identify the vector
bundles occurring in the complex, (11.5), with appropriate
tensor powers of the tautology bundle, V. If we do so, this
complex takes the form.

$$(11.6) \quad 0 \longrightarrow \Gamma(S^2(V^*)) \xrightarrow{\ \kappa\ } \Gamma(S^4(V) \otimes (\Lambda^2[V]^*)^2)$$
$$\xrightarrow{\ \sigma\ } \Gamma(S^*(V) \otimes \Lambda^2[V]) \xrightarrow{\ \beta\ } \Gamma(S^2(V) \otimes (\Lambda^2[V])^3) \longrightarrow 0 \ .$$

By the Gasqui-Goldschmidt theorem, this complex computes the
cohomology of $M_{2,1}$ with values in the sheaf of conformal
vector fields. However, for $M_{2,1}$ this sheaf is the constant
sheaf, **sp(2,ℝ)**. Therefore, we can immediately write down its
cohomology groups:

Proposition 11.3 The cohomology groups of (11.6) are just the
DeRham cohomology groups of M tensored by $sp(2,ℝ)$; i.e.,
$H^0 = H^1 = sp(2,ℝ)$ and $H^2 = H^3 = 0$.

Now let us return to the Casian diagram associated with χ_1. Notice that the lattice depicted in this diagram is generated by the entries which we've labelled A, B, C, D and $\bar{D}$; so we only have to identify these entries with subrepresentations of $\text{ind}_G \chi_1$. By experimenting with various possibilities it is easy to convince oneself that there is *only one way* to choose A, B, C, D and $\bar{D}$ so as to obtain the configuration in figure one. As we have pointed out already, A has to be the pre-image of the unique proper subrepresentation of $\text{ind}_G \chi_2$. As for B and C, B has to be the kernel of σ and C the image of κ. (In view of Proposition 11.3, this accounts for the slashed arrow joining B to C.) Finally D and $\bar{D}$ have to be the (boundary data of) holomorphic and anti-holomorphic sections of (9.5) over $M_{2,1}^+$. It is clear, in fact, that these choices are consistent with figure one; and it is not hard to see that one can only obtain this figure by positioning A, B, C, D and $\bar{D}$ as we've indicated.

Notice by the way that if we truncate Casian's lattice by throwing away everything below A_{bis} we obtain the lattice for χ_2, which is consistent with Theorem 10.1 since A_{bis} corresponds to the kernel of R. By throwing away everything above A_{bis} we obtain the lattice structure of the kernel of R itself. In particular we obtain

Proposition 11.4.   The kernel of R is spanned by the image
of κ and by the (distributional) boundary data of holomorphic
and anti-holomorphic sections of (9.5) over $M_{2,1}^+$.

We have already pointed out in section nine that the
space of infinitesmal conformal deformations of $M_{2,1}$ is *not*,
properly speaking, the space of sections of (9.5), but rather
the quotient of this space by the image of κ; i.e.,

$$(11.7) \qquad def(M_{2,1}) = \Gamma(S^4(V) \otimes \Lambda^2[V^*])^2)/\kappa(\Gamma(S^2(V^*))).$$

We obtain the lattice structure of this space by throwing away
everything in figure one below the position C. In particular
we obtain:

Proposition: The space, $def(M_{2,1})$, contains three irreducible
pieces, the pieces associated with the holomorphic and anti-
holomorphic sections of (9.5) over $M_{2,1}^+$ and a piece isomorphic to $sp(2,\mathbb{C})$. The direct sum of these three pieces is a
(unique) maximal invariant subspace of $def(M_{2,1})$.

By the Gasqui-Goldschmidt theorem the $sp(2,\mathbb{C})$ piece
corresponds to conformally flat deformations of $M_{2,1}$, and we
already know that the sum of the holomorphic and anti-
holomorphic pieces correspond to cyclic deformations of $M_{2,1}$;
so, to summarize the results of this section, we have proved:

Theorem 11.5. The space, $def(M_{2,1})$, possesses a unique maximal $Sp(2,\mathbb{R})$-invariant subspace. Moreover, this subspace is the direct sum of two subspaces, one being the space of infinitesmal *conformally flat* deformations of $M_{2,1}$ and the other the space of infinitesmal *cyclic* deformations of $M_{2,1}$.

Notice by the way that if we restrict ourselves to *real* sections of $S^4(V) \otimes (\Lambda^2[V^*])^{-2}$ and the corresponding *real* part of the space, (11.7), both these subspaces are irreducibles. For geometric purposes it is, of course, only the real part of $def(M_{2,1})$ that we're interested in. (Roughly speaking, our theorem says that there are only *two* interesting ways of deforming $M_{2,1}$, one by cyclic deformations and the other by conformally flat deformations!) Since these two subspaces don't intersect, we obtain the following curious result:

Corollary. There are *no* conformally flat cyclic deformations of $M_{2,1}$.

It turns out, by the way, that there are *lots* of conformally flat Zollfrei metrics on the manifold, $M_{2,1}$. In fact these exist such metrics *arbitrarily close* to the standard Einstein metric. However, the null-geodesics associated with such metrics have much larger periods than the null-geodiscis associated with the standard metric. (The analogue of this

situation is the situation in dimension two where *all* metrics
are conformally flat. For Zollfrei metrics null-geodesic flow
is described by vector fields of the form, $\Xi_\alpha = \partial/\partial x + \alpha \, \partial/\partial y$,
α rational, on the two-torus, $\mathbb{R}^2/\mathbb{Z}^2$. For Zollfrei metrics
close to the standard metric, this rational number is close to
zero; so the null-geodesics have to spiral many times around
the torus before they close up.)

Another consequence of Theorem 11.5 is the non—existence
of *isochronous* cyclic deformations of $M_{2,1}$. We recall that
the maximal compact subgroup of $Sp(2,\mathbb{R})$ is $U(2)$ and that
its center, S^1, is the group of time displacements of $M_{2,1}$.
A deformation of $M_{2,1}$ is isochronous if it is invariant with
respect to time displacements. We will prove

Theorem 11.6.   There are no isochronous cyclic deformations of
$M_{2,1}$.

Proof. The Siegel upper half-space, $M_{2,1}^+$, can, by means of
the Cayley transform, be identified with the generalized unit
disk

$$D_3 = \{Z = \begin{bmatrix} z_1, & z_2 \\ z_2, & z_3 \end{bmatrix}, \ z_i \in \mathbb{C}, \ Z\bar{Z} \leq I\}.$$

(See [39].) $U(2)$ acts on D_3 by the linear action, $A \rightarrow L_A$,
where

(11.8) $L_A Z = AZA^t$;

so in particular, the group of time displacements acts on D_3 by the action:

(11.9) $e^{it} \longrightarrow$ multiplication by e^{2it}.

Suppose now that there exists a symmetric contravariant two-tensor

(11.10) $\sum f_{ij}(z)dz_i dz_j$, $1 \leq i,j \leq 3$,

which is both holomorphic (i.e., the f_{ij}'s are holomorphic functions on D_3) and invariant with respect to the transformations (11.9). Invariance with respect to (11.9) means that

(11.11) $e^{2it}f_{ij}(e^{it}z) = f_{ij}(z)$.

This identity holds, by assumption, for all real t; hence, since the f_{ij}'s are holomorphic in z, it holds for all complex values of t for which the left hand side of (11.11) makes sense. In particular

(11.12) $\lambda^3 f_{ij}(\lambda z) = f_{ij}(z)$

for all $\lambda \in \mathbb{R}$ for which both sides of (11.12) make sense.
Setting $\lambda = 0$ we see that $f_{ij} = 0$.

§12. We will next show that the infinitesmal cyclic deformations of $M_{2,1}$ can in some sense be thought of as "free gravitons". To start with consider the holomorphic analogue of the
complex (11.6): Let E_1, E_2, E_3 and E_4 be the vector
bundles, $S^2(V^*)$, $S^4(V) \otimes (\Lambda^2[V^*])^2$, $S^4(V) \otimes \Lambda^2[V]$ and
$S^2(V) \otimes (\Lambda^2[V])^3$ respectively, and let $\Gamma_{hol}(E_i)$ be the space
of K-finite holomorphic sections of E_i over $M_{2,1}^+$. The complex, (11.6), has the holomorphic counterpart

$$(12.1) \qquad 0 \longrightarrow \Gamma_{hol}(E_1) \xrightarrow{\kappa} \Gamma_{hol}(E_2) \xrightarrow{\sigma} \Gamma_{hol}(E_3)$$
$$\xrightarrow{\beta} \Gamma_{hol}(E_4) \longrightarrow 0,$$

κ, σ and β being the holomorphic versions of the κ, σ and
β of (11.6).

Proposition 12.1. The sequence (12.1) is exact.

Proof. We can identify $M_{2,1}^+$ with the bounded Siegel domain,
D_3, in $\mathbb{C}^3$ by means of the Cayley transform (see §11). Let
$J_{hol}^{\kappa}(E_i)_0$ be the space of k–jets of holomorphic sections of
E_i at the origin in D_3. First we will prove

Lemma. The mapping, $\Gamma_{hol}(E_i) \longrightarrow J_{hol}^{\kappa}(E_i)_0$, is surjective.

<u>Proof</u>. The tautology bundle, V, is holomorphically trivial over $M^+_{2,1}$; hence, so is the bundle E_i. Therefore we can think of a section of E_i as being an N-tuple of holomorphic functions, where N is the fiber dimension of E_i. Therefore, it is enough to prove that there exists a holomorphic functions on D_3 having a pre-assigned k-jet at 0. This is, obvious, however; in fact we can choose this holomorphic function to be a polynomial in the linear coordinates z_1, z_2 and z_3. (See (11.8).)   Q.E.D.

By [10] the formal sequence

$$(12.2) \qquad 0 \longrightarrow J^\infty_{hol}(E_1)_0 \xrightarrow{\;\kappa\;} J^\infty_{hol}(E_2)_0 \xrightarrow{\;\sigma\;} J^\infty_{hol}(E_3)_0$$
$$\xrightarrow{\;\beta\;} J^\infty_{hol}(E_4)_0 \longrightarrow 0$$

is exact. Moreover, the map which assigns to each holomorphic section, s, of E_i its formal jet, $J^\infty(S)_0$, defines an injection morphism of the complex (12.1) into the complex (12.2). Now decompose $J^\infty_{hol}(E_i)_0$ into its K-types (K being the maximal compact subgroup, U(2), of Sp(2,$\mathbb{R}$) which stabilizes the origin in D_3.) By the lemma, the morphism of (12.1) into (12.2) is an isomorphism at each "K-level"; therefore since (12.2) is exact so is (12.1).   Q.E.D.

It follows from the proposition that the cokernel of κ in (12.1), which is, by Theorem 11.5, the space of infinitesmal cyclic deformations of $M_{2,1}$, gets imbedded by σ into $\Gamma_{hol}(S^4(V) \otimes \Lambda^2[V])$ as the *kernel* of β; i.e., there is a canonical identification

(12.3)   inf. cylic deformations of $M_{2,1}$

$\cong$ the kernel of β in $\Gamma_{hol}(S^4(V) \otimes \Lambda^2[V])$.

We will now show that the right hand side has a very simple interpretation in terms of four-dimensional Minkowskian geometry. First, however, lets briefly review the definition of $M_{2,1}^+$ and remind our readers how the *four-dimensional* analogue, $M_{3,1}^+$, of $M_{2,1}^+$ is defined. The standard symplectic form, ω, on $\mathbb{R}^4$ has a $\mathbb{C}$-linear extension, $\omega_{\mathbb{C}}$, to $\mathbb{C}^4$; and associated with this extension is the Hermetian form

(12.4) $H(v,\bar{v}) = (1/\sqrt{-1})\omega_{\mathbb{C}}(v,\bar{v})$

of signature $(2,2)$. Let $SU(2,2)$ be the group of complex linear transformations of $\mathbb{C}^4$ which preserve H and have determinant one. It is clear from (12.4) that $SU(2,2)$ contains $Sp(2,\mathbb{R})$. On the other hand $SU(2,2)$ is the group of conformal symmetries of compactified Minkowski 4–space; and, in fact, one way of seeing this is to note that compactified Minkowski 4-space has an *intrinsic* description in terms of the data above: By definition $M_{2,1}^+$ is the Grassmannian of two—

dimensional Lagrangian subspaces of $\mathbb{C}^4$ on which $H > 0$.
Consider the slightly larger Grassmannian: *all* two-dimensional
subspaces of $\mathbb{C}^4$ on which $H > 0$. This Grassmannian is a
four–dimensional homogeneous complex domain (with $SU(2,2)$ as
its group of automorphisms); and it turns out that it bears
exactly the same relation to $M_{3,1}$ that $M_{2,1}^+$ does to $M_{2,1}$:
namely its Shilov boundary is $M_{3,1}$. Henceforth, we will
denote this domain by $M_{3,1}^+$. Notice by the way that there is a
natural imbedding:

$$(12.5) \qquad\qquad \iota: M_{2,1}^+ \longrightarrow M_{3,1}^+.$$

Living on $M_{3,1}^+$ are the "free mass-zero spin-k/2 particles" of elementary particle physics, which are, by definition,
certain irreducible representations of $SU(2,2)$. We will
briefly describe how these representations are defined: Let V
be the "tautology bundle" of $M_{3,1}^+$. (This is defined exactly
as for $M_{2,1}^+$, namely its fiber at p is the two-dimensional
subspace of $\mathbb{C}^4$ represented by p.) It is clear from this
definition that V is a vector subbundle of the trivial
bundle, $\mathbb{C}^4$. Let W be the quotient bundle, $\underline{\mathbb{C}}^4/V$. It turns
out that for all integers, $k > 0$, there is an $SU(2,2)$–
invariant first order differential operator, ∂_k, whose domain
is the space of holomorphic sections of the vector bundle

(12.6) $$S^k(V) \otimes \Lambda^2[V],$$

and whose range is the space of holomorphic sections of the vector bundle

(12.7) $$S^{k-1}(V) \otimes (\Lambda^2[V])^2 \otimes W^*.$$

(For instance for $k = 1$, ∂_k is the usual Dirac operator. ∂_k is sometimes called the "spin–$k/2$" Dirac operator). The "free mass–zero spin–$k/2$ fields" are defined to be the holomorphic sections, s, of (12.6) which satisfy

(12.8) $$\partial_k s = 0.$$

(See [21], page 85, or [7].)

Notice that except for $k = 1$ the fiber dimension of (12.6) is less than the fiber dimension of (12.7); so the equation (12.8) is over-determined. Therefore, if M is a non-characteristic hypersurface in $M_{3,1}^+$ and one wants to solve (12.8) with pre-assigned initial data, $s = s_M$ on M, s_M has to satisfy a "constraint equation." We will describe this constraint equation for the imbedding (12.5): Over $M_{2,1}^+$ there is an isomorphism of bundles $V \cong W^*$ given by the symplectic structure on $\mathbb{C}^4$; so the bundle (12.7) is isomorphic to $S^{k-1}(V) \otimes (\Lambda^2[V])^2 \otimes V$, which is also isomorphic to

(12.9) $$S^{k-1}(V) \otimes (\Lambda^2[V])^3 \otimes V^* .$$

By contracting the first and third terms in this tensor product one gets a morphism, γ, of the bundle, (12.9), onto the bundle

(12.10) $$S^{k-2}(V) \otimes (\Lambda^2[V])^3.$$

It is not hard to show that at every $p \in M^+_{2,1}$ the kernel of γ is the image of the symbol map, $\sigma(\partial_k)(\xi)$, for the conormal vector, ξ, to $M^+_{2,1}$ at p. Therefore, if s is a section of $S^k(V) \otimes \Lambda^2[V]$ defined on some subset, U, of $M^+_{2,1}$ and $\tilde{s}$ is a section of $S^k(V) \otimes \Lambda^2[V]$ defined on some open subset, $\tilde{U}$, of $M^+_{3,1}$ containing U, and equal to s on U, the expression

(12.11) $$\gamma(\partial_k \tilde{s}|U)$$

is independent of the choice of $\tilde{s}$ (i.e., depends only on s). Hence (12.11) defines a first-order differential operator,

(12.12) $$\partial^b_k: \Gamma(S^k(V) \otimes \Lambda^2[V]) \longrightarrow \Gamma(S^{k-2}(V) \otimes (\Lambda^2[V])^3),$$

which is just the "tangential component" of ∂_k. In particular, the constraint equation for $M^+_{2,1}$ is

(12.13) $$\partial^b_k s = 0.$$

Now it is not hard to show that, for $k = 4$, the operator

$$\partial_4^b: \Gamma_{hol}(S^4(V) \otimes \Lambda^2[V] \rightarrow \Gamma_{hol}(S^4(V) \otimes \Lambda^2[V])^3)$$

is identical with the operator, β, in the sequence (12.1). Therefore, the restriction mapping

$$(12.14) \quad \iota: \Gamma_{hol}(S^4(V) \otimes \Lambda^2[V], M_{3,1}^+) \rightarrow \Gamma_{hol}(S^4(V) \otimes \Lambda^2[V], M_{2,1}^+)$$

maps the kernel of ∂_k into the kernel of β. Since $M_{2,1}^+$ is non-characteristic, the restriction of a holomorphic section of (12.6) which satisfies $\partial_k s = 0$ can vanish identically on $M_{2,1}^+$ if and only if all the terms in its Taylor series expansion in the direction normal to $M_{2,1}^+$ vanish identically on $M_{2,1}^+$; and since s is holomorphic, this means that s itself has to vanish. On the other hand we know from §11 that the space (12.3) is irreducible. Therefore, ι^* maps the space of solutions of (12.8) *surjectively* into the space (12.3). Thus we've proved:

Theorem. The space of solutions of the free mass–zero spin 2 field equations on $M_{3,1}^+$ is isomorphic to the space (12.3). In other words

---
Infinitesmal cyclic deformations of $M_{2,1}$= "free gravitons"

     We will conclude this section with a few comments about this result:

     1.   We are uncertain whether this result is just a piece of mathematical serendipity or whether it has interesting physical ramifications. For instance, do cyclic models in (2+1)-dimensions have anything to do with solutions of Einstein's equations in (3+1)-dimensions? For some speculations on how the two may be related, see the comments at the end of §15.

     2.   One consequence of this result is that the space of infinitesmal cyclic deformations of $M_{2,1}$ has a *natural complex structure*. This also may just be a piece of mathematical serendipity, or it may mean that there are twistorial aspects to the theory of cyclic models analogous to the "Penrose picture" of GR in dimension four.

3.   One interesting consequence of the theorem is that
the space, (12.3), has an SU(2,2)-invariant inner product
defined on it. (This is because *all* the mass—zero spin-k/2
representations are unitarizable. See [21].) There are a
number of useful applications of this fact, and we will briefly
describe one of these below:

Let us denote by $\mathcal{M}_{\text{cyclic}}$ the set of all conformal
classes of Zollfrei metrics on $M_{2,1}$, and let 0 be the base-
point in $\mathcal{M}_{\text{cyclic}}$ represented by the standard Einstein metric.
In §18 we will define, for each Zollfrei metric, g, a collec-
tion of conformal "Floquet" invariants. One can think of each
of these invariants as defining a function

$$F: \mathcal{M}_{\text{cyclic}} \longrightarrow \mathbb{R}.$$

We claim that, *without knowing anything of the specifics of how
this function is defined*, one can see at once that

$$(\delta F)_0 = 0$$

and

$$(\delta^2 F)_0 = C_F(\cdot, \cdot)$$

where C_F is a real constant and the quadratic form on the
right is the real part of the Hermitian inner product on the

Hilbert space, $\mathbb{H}$, of mass-zero spin 2 fields. Indeed $\delta_0 F$ is an $Sp(2,\mathbb{R})$-invariant linear form on this space so it has to be zero because of irreducibility, and $(\delta^2 F)_0$ has to have the form indicated above for the same reason. In particular either

      $i.$ F has a strict local maximum at 0

              or

     $ii.$ F has a strict local minimum at 0

              or

   $iii.$ $(\delta^2 F)_0 \equiv 0.$

One of the most accessible of the Floquet invariants is the *symbolic trace* of the Floquet operator. (See (18.2).) For this invariant, $(\delta^2 F)_0 \equiv 0$. In principal one should be able to compute $(\delta^3 F)_0$ by decomposing the Hermetian tensor product

$$\mathbb{H} \underset{s}{\otimes} \bar{\mathbb{H}}$$

into $Sp(2,\mathbb{R})$ irreducibles. (For the analogous problem for $SU(2,2)$, see [22].)

§13. In this section we will discuss the results of §10–11
from a slightly different slant. First, however, we will say a
few words about "gauge–fixing" in (2+1)-dimensional general
relativity: Let $\mathcal{M}_{2,1}$ be the set of all Lorentz metrics on
$M_{2,1}$. The group

$$\mathcal{G} = \text{Diff}(M_{2,1}) \ltimes C^{\infty}(M_{2,1})^{+}$$

acts on $\mathcal{M}_{2,1}$ by "conformal change of variables", and any two
elements of $\mathcal{M}_{2,1}$ which are on the same $\mathcal{G}$-orbit are identical
as far as their causal properties are concerned. Therefore, it
is tempting to try to find a global cross-section to the action
of $\mathcal{G}$ on $\mathcal{M}_{2,1}$ (i.e., a "gauge") so that the basic properties
of (2+1)-dimensional conformal spaces become particularly
simple "in this gauge." It is probably impossible to do this;
however, what we will show is that, for every metric which is
C^{0}-close to the standard Einstein metric, its $\mathcal{G}$-orbit contains
a metric with a particularly simple canonical form. This form
is easiest to describe for the double cover of $M_{2,1}$, i.e.,
for $S^2 \times S^1$, so we will start with metrics on it.

        For each $a \in S^1$ let $\iota_a \colon S^2 \to S^2 \times S^1$ be the inclusion
mapping, $x \to (x,a)$, and let $(d\sigma)^2$ be a Lorentzian metric on
$S^2 \times S^1$ which has the property that for every $a \in S^1$, $\iota_a^*(d\sigma)^2$
is Riemannian. (In other words for every $a \in S^1$, the surface,
$S^2 \times \{a\}$, in $S^2 \times S^1$ is *space-like* with respect to $(d\sigma)^2$.) The

Einstein metric on $S^2 \times S^1$, $(dx)^2 - (dt)^2$, obviously has this property, and so does any Lorentzian metric which is C^0 close to the Einstein metric.

Proposition 13.1. There exists a diffeomorphism, ϕ, of $S^2 \times S^1$ onto itself and a smooth positive function, ρ, on $S^2 \times S^1$ such that for all $a \in S^1$

(13.1) $\iota_a^* \rho \phi^* (d\sigma)^2 = (dx)^2,$

$(dx)^2$ being the standard $SO(3)$-invariant metric on S^2.

Proof. Let
$$p_i : S^1 \rightarrow S^2, \qquad i = 1,2,3,$$

be a smooth mapping with the property that for all $a \in S^1$ the points, $p_1(a)$, $p_2(a)$ and $p_3(a)$ are distinct. By the Korn-Lichtenstein theorem there exists, for each $a \in S^1$, a diffeomorphism, ϕ_a, of S^2 onto itself and a smooth function, γ_a, on S^2 such that

$$\phi_a^* \iota_a^* (d\sigma)^2 = \gamma_a (dx)^2.$$

Moreover, if we require that the $p_i(a)$'s be fixed points of ϕ_a, then ϕ_a and γ_a will be unique. (In particular, they

will depend smoothly on the parameter, a.) Now, for all
$a \in S^1$, set $\phi = \phi_a$ and $\rho = \gamma_a^{-1}$ on the fiber of $S^2 \times S^1$
above a.   Q.E.D.

Next let $(d\sigma)^2$ be a Lorentzian metric on $M_{2,1}$. Let π
be the covering map of $S^2 \times S^1$ onto $M_{2,1}$ and assume as above
that, for all $a \in S^1$, $\iota_a^* \pi^* (d\sigma)^2$ is Riemannian.

<u>Proposition 13.2</u>.   There exists a diffeomorphism, ϕ, of $M_{2,1}$
onto itself and a smooth function, ρ, on $M_{2,1}$ such that for
all $a \in S^1$

(13.2) $\iota_a^* \pi^* \rho \phi^* (d\sigma)^2 = (dx)^2.$

<u>Proof</u>.   We can think of $(d\sigma)^2$ as a Lorentzian metric on
$S^2 \times S^1$ which is invariant with respect to the involution

(13.3) $\kappa: S^2 \times S^1 \longrightarrow S^2 \times S^1,$ $\kappa(x,t) = (-x, t+\pi).$

Now in the proof of Proposition 13.1 choose the p_i's such that
for all values of t the involution, (13.3), maps the set of
$p_i(t)$'s onto the set of $p_i(t + \pi)$'s. For example, set

$$p_1(t) = (1,0,0),$$
$$p_2(t) = (-1,0,0),$$
$$p_3(t) = (0, \cos t, \sin t).$$

Then the ϕ and ρ constructed in Proposition 13.1 will be invariant with respect to (13.3), and hence ϕ can be regarded as a diffeomorphism of $M_{2,1}$ and ρ as a function on $M_{2,1}$. Q.E.D.

Suppose now that $(d\sigma)^2$ has been normalized as described in Proposition 13.1, i.e., has the property that for all $a \in S^1$, $\iota_a^*(d\sigma)^2 = (dx)^2$. This implies that there exists a one-form, $\omega \in \Omega^1(S^2 \times S^1)$, such that

$$(13.4) \qquad\qquad (d\sigma)^2 = (dx)^2 - \omega \circ dt .$$

We can write this one form in the form

$$(13.5) \qquad\qquad \mu - f\ dt + \nu$$

where ν involves no dt term, i.e.,

$$(13.6) \qquad\qquad \iota(\tfrac{\partial}{\partial t})\nu = 0.$$

In standard relativistic parlance, f is called the *lapse* function of the metric $(d\sigma)^2$ and ν the *shift* vector. If $\nu = 0$ then one says that $(d\sigma)^2$ is obtained from the standard Einstein metric by a *lapse transformation*. This means that of each point, p, in $S^2 \times S^1$, the light cone associated with

$(d\sigma)^2$ and the light cone associated with the Einstein metric
have the same axis of symmetry; but the new light cone is
elongated by a change in the velocity of light at p by a
factor, $\sqrt{f}$. (See the figure below.)

standard light cone   new light cone

On the other hand, if f = 1 one says that $(d\sigma)^2$ is
obtained from the standard Einstein metric by a *shift* transformation. In particular suppose ν is small. Then, if one
ignores effects of order $\|\nu\|^2$, a shift transformation has the
following effect on the geometry of $S^2 \times S^1$ at a fixed point,
p: The new light cone at p is congruent to the standard
light cone, but its axis of symmetry now lies in the plane
spanned by ν and the old axis of symmetry and makes an angle
of $\|\nu\|/4$ radians with the old axis of symmetry, as in the
figure below:

standard light cone   new light cone

Consider now a deformation $\{g_s, -\epsilon < s < \epsilon\}$ of the standard Einstein metric, $g_0 = (dx)^2 - (dt)^2$ on $M_{2,1}$. By Theorem (13.1) we can put this deformation into the normal form:

(13.7) $(dx)^2 - \omega_s \circ dt,$ $-\epsilon < s < \epsilon,$

ω_s being a one-form depending smoothly on s with $\omega_s = dt$. Suppose that the metrics, g_s, are all Zollfrei metrics. Then $\dot{g}_0$ has to satisfy the condition (7.2), and it is easy to see that this is equivalent to the condition

(13.8) $\int_\gamma \dot{\omega}_0 = 0$

for all null-geodesics, γ. In other words "in the temporal gauge", described in Theorem 13.1, the infinitesmal cyclic deformations of $M_{2,1}$ are just the solutions of (13.8). Its clear that *exact* one-forms satisfy (13.8); however, if $\dot{\omega}_0$ is an exact one—form, i.e., $\dot{\omega}_0 = df$, then

$$\dot{\omega}_0 \circ dt = D_\xi g_0, \qquad \text{with } \xi = f \frac{\partial}{\partial t} ;$$

so infinitesmal deformations of this type are *trivial*. However, one can obtain non-exact one forms satisfying (13.8) by
exploiting the fact that $M_{2,1}$ is the Shilov boundary of $M_{2,1}^{+}$
as we did in $\S 9$. Let ω be a holomorphic one-form on $M_{2,1}^{+}$
and let μ be the real part of its hyperfunction trace. If μ
is smooth, one can replace the integral of μ over γ in
(13.8) by the real part of the integral of ω over a contour
in $M_{2,1}^{+}$. By shrinking this contour to zero, we obtain (13.8).
It is also not hard to prove the converse of this remark using
Theorem 11.5. Namely, if $\dot{\omega}_0$ is a one-form satisfying (13.8)
then there exists an exact one-form, μ, and a holomorphic
one-form, ν, on $M_{2,1}^{+}$ such that

(13.9) $\dot{\omega}_0 = \mu + \text{Re}(\nu_0)$

ν_0 being the hyperfunction trace of ν.

So far in this article we have said very little about
Zollfrei deformations of the double cover, $S^2 \times S^1$, of $M_{2,1}$.
We will conclude this section with a few comments about $S^2 \times S^1$:
Just as for $M_{2,1}$ itself, cyclic deformations of the double
cover correspond (in the "temporal gauge") to one-forms on
$S^2 \times S^1$ which satisfy (13.8). Let κ be the involution, (13.3),
and let π, as above, be the covering map of $S^2 \times S^1$ onto $M_{2,1}$.
We will prove

Theorem 13. Every one-form on $S^2 \times S^1$ which satisfies (13.8) can be written in the form, $\mu_1 + \pi^* \mu_2$, where μ_2 is a one-form on $M_{2,1}$ of the type (13.9) and μ_1 satisfies $\kappa^* \mu_1 = -\mu_1$.

Proof. Every one form on $S^2 \times S^1$ can be written as the sum of an odd one-form (i.e., $\kappa^* \mu_1 = -\mu_1$) and an even one-form (i.e., $\kappa^* \mu_2 = \mu_2$). Moreover, if μ_2 is even it is the pull-back of a one-form on $M_{2,1}$. Therefore, it is enough to show that if μ_1 is odd, it has to satisfy (13.8). Because of the fact that $Sp(2, \mathbb{R})$ acts transitively on the space of null-geodesics it is enough to check (13.8) on the null-geodesic, $\gamma(t) = (\cos t, \sin t, 0, t)$, $0 \leq t \leq 2\pi$. Note that $\kappa \circ \gamma(t) = \gamma(t+\pi)$; so if $\kappa^* \mu_1 = -\mu_1$, $\gamma^* \mu_1$ has to be of the form, $f(t)dt$, with $f(t+\pi) = -f(t)$ thus

$$\int_0^{2\pi} f(t)dt = 0.$$

Q.E.D.

We know from §11 that there are no *isochronous* cyclic deformations of $M_{2,1}$. The situation is quite different on the double cover. In fact notice that on $S^2 \times S^1$ all isochronous (S^1-invariant) one-forms can be written in the form

$$f_1(x)\ dx_1 + f_2(x)\ dx_2 + f_3(x)\ dx_3 - \rho(x)\ dt$$

where x_1, x_2 and x_3 are the restrictions to S^2 of the standard coordinate functions on $\mathbb{R}^3$. Moreover, by the S^2 Hodge theorem, we can write the first summand in the form, $df + *dg$, where f and g are smooth functions on S^2 and "$*$" is the S^2-Hodge star operator. Since we are not interested in trivial infinitesmal deformations, we can throw away df and consider only S^1-invariant one-forms on $S^2 \times S^1$ of the form

$$(13.9) \qquad \nu = *dg + \rho dt, \qquad \text{with } g, \rho \in C^{\infty}(S^2).$$

Now notice that the one-form (13.9) is odd if ρ is odd (i.e., $\rho(-x) = -\rho(x)$) and g is even. (The reason for the latter is that the involution, $x \rightarrow -x$, *anti*-commutes with the star operator.) Thus we conclude: There are two types of infinitesmal isochronous cyclic deformations of $S^2 \times S^1$: *odd deformations of lapse type*: $\nu = \rho \, dt$, ρ an odd function on S^2, and *even deformations of shift type*: $\nu = *dg$, g an even function on S^2.

Remark: The deformations of shift type seem not to have been known before. On the other hand those of lapse type are just the Zoll-deformations of S^2 discovered by Funk in 1913.

PART IV
THE GENERALIZED X-RAY TRANSFORM

§14. To make the convergence scheme described in §8 actually
converge we need Sobolev estimates of the form (8.12). These
estimates are needed to establish *surjectivity* of the x-ray
transform, but at some stage we will be forced also to worry
about the fact that the x-ray transform fails to be *injective*.
(For a Zollfrei manifold, M, this question is interesting for
other reasons besides: the kernel of the x-ray transform is
the space of infinitesimal cyclic deformations of M.) For
$M_{2,1}$, i.e. for the standard x-ray transform, this kernel
consists of holomorphic sections of the vector bundle, (9.5)
over $M_{2,1}^+$; but, in general, the strongest statement we will be
able to make is that something of this nature is true microlocally; the kernel consists of microfunction sections of an
appropriate vector bundle over the region of positive energy in
T*M-0. At the boundary of this region (i.e. on the light cone)
this microlocal picture becomes somewhat hazy; and, not sur-
prisingly, this is where the main analytic complications arise.
To deal with these complications we will have to "blow up" the
microlocal support of the x-ray transform along this singular

locus. The details will be spelled out in the next three
sections.

This section will be a review of standard material on
double fibrations and integral transforms. Recall that a
double fibration is a diagram of the form:

(14.1)

$$\begin{array}{ccc} & Z & \\ \pi \swarrow & & \searrow \rho \\ X & & Y \end{array}$$

where X, Y and Z are smooth manifolds and π and ρ fiber
mappings. The product of π and ρ is a mapping

(14.2) $\pi \times \rho: Z \to X \times Y$.

Definition 14.1. The diagram (14.1) is called a *double
fibration* if (14.2) is a proper differentiable imbedding of Z
into X×Y.

Notice that if y ∈ Y, then (14.2) imbeds the fiber above
y,

(14.3) $F_y = \rho^{-1}(y)$,

into X×Y as a submanifold of X×{y} (i.e., in effect, as a
submanifold of X). In other words a double fibration can be

viewed as a scheme for indexing a collection of submanifolds of
X by the points of Y. This is, of course, a symmetric situa—
tion. If $x \in X$, the fiber,

(14.4) $G_x = \pi^{-1}(x)$,

is imbedded by (14.2) as a submanifold of Y. The set, Z, is
called the *incidence relation* defined by (14.1).

Notice that

(14.5) $y \in G_x \Leftrightarrow x$ and y are incident $\Leftrightarrow x \in F_y$.

Since (14.2) is an imbedding of Z into X×Y we can
henceforth think of Z as a submanifold of X×Y. Let Γ be
the punctured conormal bundle of Z in X×Y. It is easy to
see that Γ sits inside the product of the punctured cotangent
bundles of X and Y; so there is a diagram

(14.6) $$\begin{array}{ccc} & \Gamma & \\ \pi_1 \swarrow & & \searrow \rho_1 \\ T^*X{-}0 & & T^*Y{-}0 \end{array}$$

which we will think of as the microlocal analogue of (14.1).
Notice that T*X-0 and T*Y-0 are symplectic manifolds and Γ
a *canonical* relation between them.

The incidence relations, (14.5), also have a microlocal analogue, namely:

$$(14.7) \qquad (x,\xi,y,\eta) \in \Gamma \Leftrightarrow \xi \in N_x^* F_y \Leftrightarrow \eta \in N_y^* G_x \ .$$

In fact, from (14.7) one gets a bijective linear mapping

$$(14.8) \qquad\qquad N_x^* F_y \cong N_y^* G_x$$

which we will make use of below.

There are various ways to associate integral transforms with the double fibration (14.1). The most familiar way is to equip Z with a smooth non-vanishing density. The choice of such a density (say μ) enables one to define a transform

$$(14.9) \qquad\qquad R_\mu : C_0^\infty(X) \to C^\infty(|\Lambda|Y)$$

by setting

$$(14.10) \qquad\qquad R_\mu f = \rho_*(\pi^* f \mu) \ .$$

The Schwartz kernel of this transform is the delta-function, δ_μ, on the submanifold, Z, of $X \times Y$; i.e.

$$(14.11) \qquad\qquad \langle \delta_\mu, F \rangle = \int_Z F d\mu$$

for $F \in C_0^\infty(X \times Y)$. Since δ_μ is a classical Fourier integral
distribution in Hörmander's sense, [20], and is supported
microlocally on the punctured conormal bundle of Z in $X \times Y$,
it follows that R_μ is a Fourier integral operator with Γ as
its underlying canonical relation.

We will next describe a slightly more complicated way of
associating integral operators with (14.1). Assume that the
fibers of ρ are oriented and that the orientation changes
consistently from fiber to fiber. Let $k = \dim Z - \dim Y$, and
let $\Omega^r(Y)$ and $\Omega_0^{r+k}(X)$ be the space of r-forms on Y and
of compactly supported $(r+k)$-forms on X. Finally let F be
a function on Z which is smooth and positive. One can define
an operator

$$(14.12) \qquad\qquad R_F : \Omega_0^{r+k}(X) \to \Omega^r(Y)$$

by setting

$$(14.13) \qquad\qquad R_F \omega = \rho_* F \pi^* \omega \ .$$

Here ρ_* is the "fiber integration" operation or, in the term-
inology of algebraic topologists, the "Thom-Gysin" morphism.
From the geometric point of view this transform is a much more
formidable object than the transform (14.9). However, micro–

locally, as we will see, these two transforms can be handled in essentially the same way.

Lets come back again to the diagram (14.6). We will mainly be interested in properties of this diagram which are invariant under symplectic changes of coordinates. To under-line this fact we will, for the moment, replace T*X-0 and T*Y-0 by arbitrary symplectic manifolds, M and N, in (14.6) and consider the general set-up

(14.14)
$$\begin{array}{ccc} & \Gamma & \\ \pi \swarrow & & \searrow \rho \\ M & & N \end{array}$$

Γ being an imbedded Lagrangian submanifold of M×N. We will henceforth assume, however, that M and N have the same dimension. With this assumption we will show

Proposition 14.1. The maps, π and ρ have the same rank at all points of Γ. In particular the critical set of π coincides with the critical set of ρ.

Proof. It is enough to prove the linear version of this proposition. Namely let V and W be symplectic vector spaces of the same dimension, and Γ a Lagrangian subspace of V×W. It's enough to prove that the space

$$\{v \in V, \; (v,w) \in \Gamma \text{ for some } w\}$$

and the space

$$(14.15) \qquad \{w \in W, \; (v,w) \in \Gamma \text{ for some } v\}$$

have the same dimension. This in turn is a consequence of the following lemma which we leave for the reader to verify

<u>Lemma 14.2</u>. The space $\{w \in W, \; (0,w) \in \Gamma\}$ is the symplectic ortho-complement in W of the space (14.15)

In the examples which we will discuss in §15, the diagram (14.14) will be equipped with an additional piece of structure, namely there will exist an involution

$$(14.16) \qquad\qquad \sigma : \Gamma \to \Gamma$$

with the property:

(14.17)i. *The mapping, ρ, in* (14.14) *is a fold and σ is the involution canonically associated with ρ.*[*]

---
[*] For a review of elementary properties of involutions and folds, see the appendix at the end of this section.

This is a strong assumption, not only about ρ, but about π as well. Before continuing, we will list a few of its consequence: First, the fixed point set, S, of σ has to be a codimension one submanifold of Γ and has to be identical with the set of critical points of ρ. Second, the corank of ρ along S has to be one (and, hence, because of proposition 14.1, S has to be the critical set of π and *its* corank along S has to be one). Third, the restriction of ρ to S has to be an *immersion* of S into N. (In particular the image of S in N, which we will henceforth denote by Σ, has to be an immersed codimension one co-isotropic submanifold of N.)

We will add to (14.17)i a complementary assumption about the mapping, π:

(14.17)ii. *For all* $s \in S$, $\ker d\pi_S \subset T_S S$.

Notice that $\ker d\pi_S$ and $\ker d\rho_S$ are both one-dimensional at $s \in S$. The characteristic property of a fold is that $\ker d\rho_S$ and $T_S S$ are *complementary* subspaces of T_S; so the assumption (14.17)ii says basically that at *no* point, $s \in S$, does π have a singularity of fold type. We will now prove a result which will give us a precise picture of what the singularities of π are like along S.

<u>Proposition 14.3</u>. Let ω_M be the symplectic form on M and $(\omega_M)^d$ the corresponding volume form (2d being the dimension of M). Then $\pi^*(\omega_M)^d$ vanishes to exactly first order on S.

<u>Proof</u>. Let ω_N be the symplectic form on N. Then $\rho^*(\omega_N)^d$ vanishes to exactly first order on S since ρ is a fold. However, since Γ is a canonical relation, $\pi^*(\omega_M)^d = \rho^*(\omega_M)^d$. Q.E.D.

Coupled with (14.17)ii, this result has the following important consequence:

<u>Proposition 14.4</u>. π is a "blowing-down" of Γ (S being the submanifold of Γ which gets "blown-down" by π).

<u>Remark</u>. See the appendix at the end of this section for an outline of the general theory of "blowing-down" and "blowing-up" mappings.

By (14.17)ii the map, π, restricted to S is of corank 2 everywhere, so its image, W is an immersed submanifold of M of codimension two and $\pi: S \to W$ is a submersion. Since ρ maps S diffeomorphically onto Σ we can identify S with Σ and think of the restriction of π to S as a fiber mapping

$$\pi: \Sigma \to W.$$

<u>Proposition 14.5</u>. W is a symplectic submanifold of M and is isomorphic, as a symplectic manifold, to the symplectic reduction of the codimension one co-isotropic manifold, Σ.

<u>Proof</u>. Like proposition 14.1 this is really a statement about *linear* symplectic spaces. Namely let V and W be 2d–dimensional symplectic vector spaces, Γ a Lagrangian subspace of V×W, S a codimension one subspace of Γ and π and ρ the projections of Γ onto V and W. Suppose that π and ρ are of corank one, and that the restriction of ρ to S is injective and the restriction of π to S has a one–dimensional kernel. What is at issue is: "Does ρ map (ker $π$)∩S bijectively onto the symplectic orthocomplement of $ρ(S)$ in W?" Notice, however, that (ker $π$) ∩ S = ker $π$ and $ρ(S)$ = $ρ(Γ)$; so this question can be rephrased. "Does ρ map ker $π$ onto the symplectic orthocomplement of $ρ(Γ)$?" This question, however, is just a rephrasing of the conclusion of lemma 14.2. Q.E.D.

<u>Remarks</u>.

1. The results we've just proved show that π and ρ have very simple singularities along S, i.e. for every point, s ∈ S there are simple local canonical forms for π and ρ near s. The analytic details in the last two sections of part

four could be enormously simplified if one could show that
there is a "local symplectic canonical form" for the diagram
(14.14) itself analogous to the canonical form for the "folding
canonical relation" described in [20], XXI. It turns out that
there is such a canonical form, but it requires an additional
hypothesis on σ: Replacing the image of Γ in M by M
itself we can, without loss of generality, assume that π is
surjective. Then π maps Γ-S diffeomorphically onto M-W;
so there is a symplectic involution of M-W onto M-W corresponding to σ. It is easy to see that this involution extends
continuously to M with W as its fixed point set. We will
say that σ is *tameable* if this involution is C^∞. If σ is
tameable, a canonical form of the type we are looking for can
be found in [27]. We won't bother to describe it here since σ
is, unfortunately, *not* tameable for most of the examples discussed in §15. (In fact the only example we know of for which
it *is* tameable is $M_{2,1}$ with its standard causal structure.)

    2. Canonical diagrams with the properties, (14.17),
occur in diffraction theory, in particular, in connection with
the diffraction of curved rays by a totally geodesic obstacle.
In these problems σ is seldom tameable; in fact the obstructions to tameness seem to be interesting symplectic invariants
of these problems.

APPENDIX I. (Folds and involutions.) Let X and Y be n-dimensional manifolds, S a codimension one submanifold of X and ω a nowhere vanishing volume form on Y.

Definition. A smooth mapping, $f: X \rightarrow Y$, is a *fold* (with S as its fold locus) if the following conditions are satisfied:

(I)

     a) S is the set of critical points of f.

     b) At every point, $s \in S$, f is of corank one and the kernel of df_s intersects $T_s S$ in $\{0\}$.

     c) $f^*\omega$ vanishes exactly to first order on S.

Remark. The third condition is obviously independent of ω. It is just an invariant way of saying that the Jacobian determinant of f vanishes to first order along S.

Theorem. If f satisfies the hypotheses I a)-c), then at every point $s \in S$ one can find a local system of coordinates $(x_1, \ldots, x_n)$ centered at s and a local system of coordinates $(y_1, \ldots, y_n)$ centered at f(s) so that $f^*y_n = x$, $i = 1, \ldots, n-1$ and $f^*y_n = x_n^2$.

Proof. Locally f maps S diffeomorphically onto a codimension one submanifold of Y. Choose coordinates $y_1, \ldots, y_n$ centered at f(s) so that the image of S is the set $y_n = 0$.

The functions $f^*y_1, \ldots, f^*y_{n-1}$ are independent at s; so we can incorporate them into a local coordinate system, $(x_1, \ldots, x_n)$; (i.e. we can arrange that $f^*y_i = x_i$ for $i \leq n-1$). Moreover, we can assume that S is defined by the equation $x_n = 0$. Let $f_n = f^*y_n$. The Jacobian determinant of f in these coordinates is $\frac{\partial f_n}{\partial x_n}$; therefore, by c), $f_n = x_n^2 h(x)$ with $h(0) \neq 0$. Replacing y_n by $-y_n$ if necessary we can assume $h(0) > 0$. Now replace x_n by $x_n\sqrt{h}$.   Q.E.D.

Let U be the coordinate patch on which the coordinate functions $(x_1, \ldots, x_n)$ are defined. Without loss of generality we can assume that if $(a_1, \ldots, a_n) \in U$ then $(a_1, \ldots, a_{n-1}, -a_n) \in U$. Let $\sigma: U \to U$ be the map: $(a_1, \ldots, a_n) \to (a_1, \ldots, -a_n)$. It is clear that a function, f, in $C^\infty(U)$ can be written in the form $g(y_1, \ldots, y_n)$ if and only if $\sigma^*f = f$; so σ is an *intrinsically defined* object.

In other words if U_1 and U_2 are two open sets of the form above and σ_1 and σ_2 the involutions on U_1 and U_2, then $\sigma_1 = \sigma_2$ on $U_1 \cap U_2$. Hence, by patching the (σ, U)'s together, we can find a neighborhood, V, of S in X and an involution, $\sigma: V \to V$, with S as fixed point set, so that in any coordinate system of the form above, σ is the involution: $(a_1, \ldots, a_n) \to (a_1, \ldots, -a_n)$. σ is called the *canonical involution* associated with the fold, f. For a more detailed discussion of folds and involutions see [20], appendix C4 to volume III.

II. (Blowing down and blowing up.) Let X, Y, S and ω be as in part I.

<u>Definition</u>. A smooth mapping $f: X \to Y$ is a *blowing-down* (with S as the submanifold of X which gets *blown down* by f) if the following conditions are satisfied

   a) S is the set of critical points of f.
   b) At every point $s \in S$ the kernel of df_s is contained in $T_s S$.
   c) f is of constant rank, k, along S and $f^*\omega$ vanishes exactly to (n-k)-th order on S.

<u>Remark</u>. As in part I the third condition is independent of the choice of ω.

<u>Theorem</u>. If f satisfies the hypotheses II a)-c), then at every point $s \in S$ one can find a local system of coordinates $(x_1,\ldots,x_n)$ centered at s and a local system of coordinates $(y_1,\ldots,y_n)$ centered at f(s) so that $f^*y_i = x_i$, $i = 1,\ldots,k$ and $f^*y_i = x_i x_1$, $i = k+1,\ldots,n$.

<u>Proof</u>. The restriction of f to S is locally a fibration of S over a codimension $(n-k+1)$ submanifold, W, of Y. Let $(y_1,\ldots,y_n)$ be a coordinate system centered at $f(s)$ such that W is the set of points where $y_1 = y_{k+1} = \cdots = y_n = 0$. Without loss of generality we can assume that $df^*y_1 \neq 0$ at s. Set $x_i = f^*y_i$, $i = 1,\ldots,k$, and extend the x's to a coordinate system, $(x_1,\ldots,x_n)$, centered at s. Note that S is defined by the equation $x_1 = 0$ in these coordinates. Since the f^*y_i's vanish on S when $i = k+1,\ldots,n$ there exist smooth functions, $g_i(x)$, $i = k+1,\ldots,n$, such that $x_1 g_i(x) = f^*y_i$. From the fact that the rank of f is k on S, we obtain $g_i = 0$ on S. Moreover, the Jacobi determinant of f in these coordinates is x_1^{n-k} times the determinant of

$$\left[\frac{\partial g_i}{\partial x_j}\right] \qquad k+1 \le i,j \le n$$

so the determinant of this matrix is non-zero on S by c). Therefore, if we make the change of coordinates

$$x_i' = x_i \ , \qquad i = 1,\ldots,k \ ,$$

and

$$x_i' = g_i(x) \ , \qquad i = k+1,\ldots,n \ ,$$

we obtain the canonical form described above.   Q.E.D.

Blowing-up is based on a global variant of this canonical form theorem. Let $f: X \rightarrow Y$ be a mapping which satisfies II a)-c) and also satisfies the following global topological hypotheses:

(II)

     d) f is proper.

     e) The image, W, of S in Y is an imbedded dimension $(k-1)$ submanifold of Y.

     f) $f: S \rightarrow W$ is a fiber mapping with *connected* fibers.

(Apropos of condition e) the conditions II a)-c) imply that W is an *immersed* dimension $(k-1)$ submanifold of Y; so condition e) simply says that W has no self—intersections. As for condition f), since $f: S \rightarrow W$ is a proper submersion, the Thom isotopy theorem implies that it is a fibration. Therefore f) simply says that the typical fiber is connected.)

Let $\mathbb{P}N$ be the projectivized normal bundle of W. By property II b), the differential of f along S gives rise to a lift of f to $\mathbb{P}N$,

(14.18)
$$
\begin{array}{ccc}
S & \xrightarrow{\;\beta_f\;} & PN \\
& {\scriptstyle f}\searrow \quad \swarrow {\scriptstyle \pi} & \\
& W &
\end{array}
$$

and, from the canonical form theorem above, one can easily see that β_f is locally a diffeomorphism.

However, since f is proper, β_f is also proper; so its
image has to be all of $\mathbb{P}N$, and β_f has to be a finite-to-one
covering. In fact even more is true: the fibers of $\mathbb{P}N$ over
W are $\mathbb{R}P^{n-k}$'s, so if n-k > 1, β_f has to be either a *dif-*
feomorphism or a *double covering*. In particular we have
proved:

Theorem. If n-k > 1, the fibers of S over W are either
(n-k)-dimensional spheres or (n-k)-dimensional real projective
spaces.

Definition. If the first alternative occurs, f is *spherical*
blowing-up of Y along W, and if the second alternative
occurs, f is a *projective blowing-up* of Y along W.

There is a simple canonical form for each of these cases.
Without loss of generality we can assume that Y is the total
space of a vector bundle

$$E \to W$$

and that W sits in Y as the zero section. Let $S = \mathbb{P}(E)$,
the projectivization of E, and let X be the set of all
triples, (p,ω,v), where p is a point of W, ω is an element of $\mathbb{P}(E_p)$ and v is a vector lying in the one–dimensional subspace of E_p corresponding to ω. The map, $\gamma: X \to Y$,
which maps (p,ω,v) to (p,v) is a projective blowing-up of

Y along W; and it is easy to see that every projective blowing–up of Y along W is isomorphic to γ in a neighborhood of W. If we replace the projective-space bundle, $\mathbb{P}(E)$, by the ray bundle, $S(E)$, we get the corresponding canonical form for the *spherical* blowing-up.

§15. Before turning to the estimates, (8.12), we will describe
again some of the ingredients that go into the formulation of
these estimates: To begin with, a *causal structure* on a compact three-manifold, M, is a rule which assigns to each $m \in M$
a convex cone

$$C_m \subset T_m \ .$$

One can always find a function, H, on T*M-0 with the property that, for all m, the dual cone

$$C_m^* \subset T_m^*$$

is just the set

$$\{\xi \in T_m^*, \ H(m,\xi) > 0\} \ .$$

Of course, H is not unique; however, the solutions of the
Hamilton-Jacobi equations

$$(15.1) \qquad\qquad \dot{x}_i = \frac{\partial H}{\partial \xi_i}, \quad \dot{\xi}_i = - \frac{\partial H}{\partial x_i}$$

for which $H(x,\xi) \equiv 0$ are the same for all choices of H (up
to parameterization). A *null-geodesic* is a curve on M which
lifts to a null-solution curve of (15.1), and M has the
"Wiederkehr property" if all its null–geodesics are periodic.
(This means, by the conventions in force in this article, that

the lifted curve is periodic and that behavior of the type
depicted in the figure below is ruled out.)

(spiraling of null-geodesics about an exceptional null-geodesic
in T*M.) In particular, for such M, the set of null-
geodesics is itself a compact three-manifold. As in §5, we
will denote this manifold by P.

For $m \in M$ let Z_m be the set of null-rays in T_m
(i.e. an element $r \in Z_m$ is a ray, $\{tv, \ t > 0\}$, with
$v \in \partial C_0$). Let

$$\pi: Z \to M$$

be the fiber bundle whose fiber at m is Z_m. Given a point,
m, and a null-ray, r, there is a unique null-geodesic, γ,
passing through m in the direction of r. Moreover, the map

$$\rho: Z \to P$$

which maps (m,r) to γ is smooth and is a fibration of P

by Z. Thus there are two natural fibration of Z, and it is
easy to see that they define a double fibration

$$(15.2) \qquad\qquad \begin{array}{c} Z \\ {}^{\pi}\diagup \quad \diagdown{}^{\rho} \\ M \qquad\quad P \end{array}$$

in the sense of §14.

Now consider the integral operator, $R_{\Psi,F}$, introduced in
§8. We claim that "built into $R_{\Psi,F}$" is an operator of the
form (14.12). To see this, fix a nowhere vanishing one-form,
μ, on M with the property that μ_m is in the interior of C_m^*
for all m ∈ M. (Such a μ exists because the C_m^*'s are con-
tractible.) Given a one-form of this type, one can define a
vector-bundle morphism

$$(15.3) \qquad\qquad \lambda_\mu : T^*M \to S^2(T^*M)$$

by tensoring fiber-wise with μ and then symmetrizing. Now
compose the operator, $R_{\Psi,F}$, with (15.3). It is easy to see
that the composite of these two operators is precisely an
operator of the form (14.12); i.e. there exists an F (dif-
ferent from the F above) such that $R_{F,\Psi} \circ \lambda_\mu$ is an operator
of the form (14.13) with r = 0 and k = 1. It is clear,
however, that if we are able to prove estimates of the form
(8.12) for $R_{F,\Psi} \circ \lambda_\mu$, we will automatically get estimates like

this for $R_{F,\Psi}$ itself. Therefore, we will jettison $R_{F,\Psi}$ and, from now on, deal with this simpler operator.

Lets turn next to the microlocal diagram

(15.4)

$$\begin{array}{ccc} & \Gamma & \\ \pi \swarrow & & \searrow \rho \\ T^*M{-}0 & & T^*P{-}0 \end{array}$$

associated with (15.2). By definition Γ is the punctured conormal bundle of Z in $X{\times}Y$; so an element of Γ is a quadruple, (m,ξ,p,η), with $m \in M$, $p \in P$, $(m,p) \in Z$, $\xi \in T_m^*{-}0$, $\eta \in T_p^*{-}0$ and (ξ,η) "conormal to Z" at (m,p). Lets see what this property means geometrically: Let γ_p be the null–geodesic indexed by p and let v be its tangent vector at m. It is clear that part of the conormality condition is $\langle v,\xi \rangle = 0$. However, this implies either that ξ is *space-like* at m or that ξ is *light-like*. If ξ is space-like, the annihilator space of ξ in T_m intersects ∂C_m in two light rays, and there are two null-geodesics, γ_{p_1} and γ_{p_2}, passing through m in the directions of these two rays. By (14.8) there exists a unique covector, $\eta_i \in T_{p_i}^*$, $i = 1,2$, such that $(m,\xi,p_i,\eta_i) \in \Gamma$. If, on the other hand, ξ is *light*-like, the annihilator space of ξ in T_m is tangent to ∂C_m along a single ray, and hence there exists a single p and $\eta \in T_p^*$ with $(m,\xi,p,\eta) \in \Gamma$. Hence we have proved:

<u>Proposition 15.1</u>. The range of $\pi: \Gamma \to$ T*M-0, is the union of
the *space-like* and *light-like* sectors in T*M-0. Over the
space-like sector π is a *two-to-one* map and over the light-
like sector it is a *one-to-one* map.

We will next prove that π is a fold. To do so it is
enough to prove that for every $m_0 \in M$, the map

$$(15.5) \qquad\qquad \pi_0: \Gamma_{m_0} \to T^*_{m_0}\text{-}0$$

is a fold (Γ_{m_0} and $T^*_{m_0}$-0 being the fiber of Γ and the
cotangent fiber above m_0). Let T_0 be the tangent space to
M at m_0, C_0 the convex cone, C_{m_0}, and Z_0 the set of
null-rays in T_0. Finally let Γ_0 be the subset of $Z_0 \times (T^*_0$-0)
consisting of all pairs, (r,ξ), with r lying in the annihilator space of ξ in T_0. It is clear that $\Gamma_0 = \Gamma_{m_0}$ and
that (15.5) is just the projection map

$$(15.5)_0 \qquad\qquad \Gamma_0 \to T^*_0\text{-}0, \qquad (r,\xi) \to \xi \; .$$

If ξ is a space-like convector, there are exactly two rays,
r_1 and r_2, with $(r_i,\xi) \in \Gamma_0$, and it is clear that if we vary
ξ in a small ball about some fixed space-like covector, ξ_0,
the r_i's vary smoothly with ξ. Thus $(15.5)_0$ is a two-to-one

covering over the space-like part of T_0^*. To show that $(15.5)_0$ is a fold, all we have to do is inspect what happens in the vicinity of a point $(r_0, \xi_0) \in \Gamma_0$ where ξ_0 is *light-like*. Since the fold property only involves the two-jet of the map $(15.5)_0$ at this point, it is enough to verify this property for the "osculating quadric" to C_0 along the ray r_0: i.e. it is enough to check the fold property in the special case where the causal structure defined by C_0 is *Lorentzian*, that is, C_0 is a quadric of signature $(2,1)$. In fact, choosing coordinates, one can assume that $T_0 = \mathbb{R}^3$ and that C_0 is defined by the quadratic inequality: $x_2^2 + x_2^2 - x_3^2 \leq 0$. However, in this case, given a space-like vector, $\xi = (\xi_1, \xi_2, 1)$, the two null-rays corresponding to ξ intersect the place $x_3 = 1$ in the points:

$$(15.6) \qquad\qquad a^\pm = v \pm tw$$

where

$$v = -\frac{(\xi_1, \xi_2)}{\xi_2^2 + \xi_2^2}, \quad w = \frac{-(\xi_2, \xi_1)}{\xi_2^2 + \xi_2^2} \qquad \text{and} \qquad t^2 = \xi_2^2 + \xi_2^2 - 1$$

From the equation (15.6) it is easy to see that $(15.5)_0$ has a fold-singularity on the set $\xi_2^2 + \xi_2^2 = 1$.   Q.E.D.

To summarize, we've proved:

Proposition 15.2. The map $\pi: \Gamma \to T^*M$-0 is a fold. Moreover, the set, S, of critical points of π gets mapped by π dif-feomorphically onto the set of light-like convectors in T^*M-0.

Next notice that the involution, $\sigma: \Gamma \to \Gamma$, associated with this fold mapping, has a very simple description in terms of the data above.

Proposition 15.3. σ preserves the fibration of Γ over M, i.e. the diagram,

$$(15.7) \qquad\qquad \Gamma \xrightarrow{\;\sigma\;} \Gamma$$
$$\searrow \quad \swarrow$$
$$M$$

commutes. Moreover, on the fiber above m, $\sigma(r_+,\xi) = (r_-,\xi)$ where r_+ and r_- are the two solutions of the equation, $\xi = 0$, in the set, Z_m, of null-rays.

(On the other hand, as we will see in §16, σ acts *in an extremely subtle way* on T^*P-0.)

Lets turn next to the map, $\rho: \Gamma \to T^*P$-0. The set of critical points of ρ is the fold set, S, of σ; and, by proposition 15.2, S is diffeomorphic to the set of *light-like*

covectors in T*M-0. As in §14 we will denote this set by Σ.
We recall (see §4) that Σ is a fiber bundle over M and that
the fiber above m has two connected components, the boundary
of the dual cone, C_m^*, and the boundary of the antipodal cone,
$-C_m^*$. Since Σ is of codimension one in T*M-0 it is co-
isotropic, and the leaves of its null-foliation are identical
with the integral curves of (15.1); i.e. we can identify
null-geodesics with leaves of the null-fibration on Σ (This
involves some redundancy, however: the null-leaf through
(m,ξ) and the null-leaf through $(m,\lambda\xi)$ correspond to the
same null-geodesic in M.) Since the null-leaves are compact
and are isotopic in Σ the null-foliation of Σ is *fibrating*:
there exists a manifold, W, and a fiber map

$$\rho_0: \Sigma \rightarrow W$$

whose fibers are the null-geodesics. As we pointed out above,
the group, $\mathbb{R}^+$, acts by homotheties on Σ and on W, and of
course this action commutes with ρ_0 Moreover, W is equipped
with a unique symplectic one-form, α_0, such that

$$(15.8) \qquad \rho_0^*\alpha_0 = \iota^*\alpha_M,$$

α_M being the standard symplectic one form on T*M and ι
inclusion. Finally notice that W is a symplectic cone: The

symplectic form, α_0, transforms in exactly the same way as the
canonical one-form, α_M, under the homothety action of $\mathbb{R}^+$.[*]
The set of "$\mathbb{R}^+$ orbits" in W is, by definition, the "base"
of this symplectic cone, and from our remark above on the
redundancies involved in the null-leaf description of the null-
geodesics, its clear that the "base" of W can be identified
with P, i.e. one has an $\mathbb{R}^+$ bundle picture:

$$(15.9) \qquad\qquad \mathbb{R}^+ \longrightarrow \begin{matrix} W \\ \downarrow \\ P \end{matrix}$$

Given a section, s, of W one can pullback α_0 by s to get
a *contact one form*, $s^*\alpha_0$, on P. Let us think of s and
$s^*\alpha_0$ as maps of P into W and T*P-0 respectively. It is
clear that there exists a (unique) $\mathbb{R}^+$ equivariant map
$\kappa: W \to T^*P$ which makes the diagram

$$\begin{matrix} W & \xrightarrow{\;\kappa\;} & T^*P\text{-}0 \\ {}_{s}\nwarrow & & \nearrow_{s^*\alpha_0} \\ & P & \end{matrix}$$

commute. We let the reader check that κ is independent of s
and is a symplectic imbedding of W into T*P-0. Moreover, it
follows immediately from the definition that

---

[*] These are all standard facts about "symplectic reduction."
The details are spelled out a little more carefully in §5.

(15.10) $\kappa^* \alpha_P = \alpha_0$

(α_P being the standard symplectic one-form on T^*P.)

Lets now return again to the diagram (15.4). We have already pointed out that π maps S diffeomorphically on Σ; so the inverse of π gives one an imbedding,

$$\kappa : \Sigma \to \Gamma \ .$$

We claim that the diagram below,

$$
\begin{array}{ccc}
\Sigma & \xrightarrow{\ \kappa\ } & \Gamma \\
\downarrow{\scriptstyle \rho_0} & & \downarrow{\scriptstyle \rho} \\
W & \longrightarrow & T^*P\text{-}0
\end{array}
$$

(15.11)

commutes.

Proof. Let (m,ξ) be a point on Σ and (m,ξ,p,η) the corresponding point on Γ. Since ξ is light-like, there is just one null-geodesic, γ, passing through m with the property that ξ is conormal to γ at m, and it is clear that this has to be the projection onto M of the null-leaf of Σ through (m,ξ); so $\gamma = \gamma_p$ and $\kappa \circ \rho_0(m,\xi) = (p,\eta_0)$ for some $\eta_0 \in T_p^*$. However, (15.08) and (15.10) imply that $\kappa^* \rho^* \alpha_p = \rho_0^* \kappa^* \alpha_p$; and from this identity, one easily concludes that $\eta = \eta_0$.   Q.E.D.

It is obvious from (15.11) that ρ satisfies the criterion (14.17)ii; so, as a corollary of proposition 14.4, we get the result:

<u>Proposition 15.4</u>. ρ is a "blowing-up" of the cotangent bundle of M along the symplectic submanifold W.

In particular, for the standard $M_{2,1}$, ρ is a "blowing-up" of projective type: ρ maps Γ-S *bijectively* onto T*P-W. (In fact twistor-theorists will recognize (15.4) as the "ambi-twistor diagram" associated with $M_{2,1}$.) The same will be true, therefore, for cyclic perturbations of $M_{2,1}$. In particular the canonical relation, Γ, will consist of two pieces,

(a) The piece, Γ-S. (This will be the graph of a canonical transformation

$$(15.12) \qquad \Phi: T*P-W \to T*M-\Sigma$$

which double-covers the "space-like" sector of T*M.)

(b) The piece, S. (This will be the graph of the "reduction" canonical relation

$$(15.13) \qquad \begin{array}{ccc} & \Sigma & \\ \rho_0 \swarrow & & \searrow \iota \\ W & & T*M \end{array}$$

mapping Σ onto W.)

A few last comments concerning the geometry of the double fibration (15.2): By definition the fibers over P in this double fibration are *null-geodesics* of M. What about the fibers over M? We claim that they also have an interesting geometric property. We have just shown that P has an intrinsic contact structure. We will prove

<u>Proposition 15.5</u>. The fibers, F_m, $m \in M$, over M in the double fibration, (15.2), are *Legendrian* submanifolds of P.

<u>Proof</u>. Let p be a point on F_m and let γ_p be the null-geodesic indexed by p. There is, up to constant multiple, a unique covector, ξ_0, in T^*_m-0 associated with the point, p; namely the covector with the property that the solution, $(m(t), \xi(t))$, of the Hamilton-Jacobi equations (15.1) having $m(0) = m$ and $\xi(0) = \xi_0$ is the lift of the null-geodesic, γ_p, to T^*M. There is also, of course, a covector, η_0, in T^*_p-0 uniquely determined up to constant multiple: the covector determining the "contact direction" at p. Moreover, by (15.10), $(m, \xi_0, p, \eta_0) \in \Gamma$; hence, under the correspondence (14.8),

$$\xi_0 \in N^*_m(\gamma_p) \longleftrightarrow \eta_0 \in N^*_p(F_m) \ .$$

In particular, $\eta_0 \in N^*_p(F_m)$.   Q.E.D.

In §12 we raised the possibility that cyclic deformations of $M_{2,1}$ might have some connection with "non-linear gravitons" in (3+1) dimensions, i.e. solutions of the Einstein field equations. Mike Eastwood suggested that we try to see this connection in terms of the twistorial approach to GR developed in [25]. The main idea in this approach is to imbed the three-parameter family of Legendrian curves described above into a four-parameter family of curves such that the "contact geometry" of this family is given by quadratic equations. Unfortunately, in the real C^∞ category, the only results this approach seems to yield at the moment are formal results of the type obtained much more effectively by Fefferman and Graham in [8], using entirely different methods. (It also, unfortunately, yields solutions of Einstein's equations in (2+2) rather than (3+1)-dimensions!)

§16. The way we will prove the estimates (8.12) will be by
showing that R_F can, after a certain amount of fine-tuning,
be used as a parametrix for R_F^t. The problem we will focus on
in this section is making sense of the composite operator,
$R_F \circ R_F^t$. We will see below that the standard composition
formulas for Fourier integral operators are not applicable to
R_F and R_F^t, because of the fact that the canonical relations,
Γ and Γ^t, don't compose cleanly. In fact we will see that
$R_F \circ R_F^t$ is *not* an F.I.O. of the standard type. For a long time
we were daunted by this problem, and we owe to Richard Melrose
the clue as to how to deal with it. Following ideas of his we
will show that the Schwartz kernel of $R_F \circ R_F^t$ is a Fourier
integral distribution on the space obtained by blowing-up P×P
along its diagonal. A consequence of this is that the
singularities introduced by the break-down of clean
intersection are, in reality, fairly innocuous. (We realized
much later, by the way, that Melrose's idea of blowing up P×P
is forced on one by the exigencies of integral geometry. If
one composes the integral transforms associated with two double
fibrations

and

one doesn't usually get another integral transform of this
type; but one can frequently get around this difficult by
blowing up an appropriate submanifold of $X_1 \times X_3$.)

Let us now turn to the details: For the purposes of this
section we can assume that our integral operators are operating
on functions and densities rather than on differential forms;
i.e. we will fix a smooth non-vanishing density, μ, on Z and
define R_μ and R_μ^t by the formulae, (14.10):

$$R_\mu f = \rho_* \mu (\pi^* f)$$

and

$$R_\mu^t g = \pi_* \mu (\rho^* g) \ .$$

Notice that R_μ maps $C^\infty(M)$ into $C^\infty(|\Lambda|P)$ and its transpose
maps $C^\infty(P)$ into $C^\infty(|\Lambda|M)$. We will also fix a non-vanishing
density, ν, on M so that we can compose R_μ with R_μ^t; i.e.
we will define the composition by the formula

(16.1) $$R_\mu \nu^{-1} R_\mu^t g = \rho_* \mu \pi^* \nu^{-1} \pi_* \mu \rho^* g \ .$$

We will first show that part of the product on the right, viz.

$$(16.2) \qquad\qquad g \rightarrow \mu \pi^* \nu^{-1} \pi_* \mu \rho^* g$$

is an F.I.0. with Γ as its underlying canonical relation. However, since π is a fiber mapping, π^* composes transversally with *any* F.I.0. which maps into $C^\infty(M)$ (and, so, in particular, with (16.3)).

The canonical relation associated with (16.2) is the subset

$$(16.4) \qquad \theta_\pi \circ \Gamma^t = \{p, \eta, z, (d\pi_2)^t \xi), \ \pi(z) = m, \ \xi \in T^*_m,$$
$$(m, \xi, p, \eta) \in \Gamma\}$$

of $(T^*P-0) \times (T^*Z-0)$, and the canonical relation associated with ρ_* is the subset

$$(16.5) \qquad\qquad \theta_\rho^t = \{(z, (d\rho_z)^t \eta, p, \eta) \ ,$$

of $(T^*Z-0) \times (T^*P-0)$.

To compose (16.4) with (16.5) we first of all have to locate all the points where the product of (16.4) and (16.5) intersects the diagonal in $(T^*Z-0) \times (T^*Z-0)$. It is clear, however, that if $(p, \eta, z, (d\pi_z)^t \xi)$ belongs to (16.4) and $(z, (d\pi_z)^t \xi, p_1, \eta_1)$ belongs to (16.5), then (m, ξ) has to be either space-like or light-like. Moreover, if (m, ξ) is space-like, either $(p_1, \eta_1) = (p, \eta)$ or $(p_1, \eta_1) = \sigma(p, \eta)$, and if (m, ξ) is light-like, then (p, η) is in W and is the image of (m, ξ) under the fiber mapping: $\Sigma \to W$. (See (15.13).)

To summarize

<u>Proposition 16.1</u>. The intersection of (16.4) and (16.5) is the union of the diagonal and of graph σ in $\Gamma \times \Gamma$, and this inter-section is non-transverse precisely on the set, diagonal $\cap$ graph σ, i.e. on the fixed point set, S, of σ.

This intersection is depicted somewhat schematically in the figure below.

(16.6)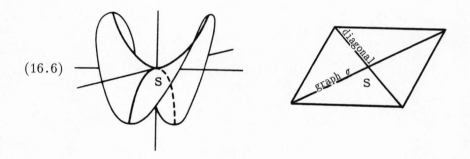

The hyperboloid is supposed to represent the set (16.4) and its tangent plane at S the set (16.5). Their intersection is pictured on the right. In the proposition, as in this picture, the hyperboloid is the graph of a function with a *non-degenerate* critical point at S. Though the intersection is

not a manifold (it has singularities) the singularities are
relatively benign: i.e. "normal crossings." Nevertheless, the
canonical relations, (16.4) and (16.5), are not composable in
the usual sense; and, in fact, the composition, regarded as a
subset of T*PxT*P, consists of three pieces (corresponding to
the three subsets pictured in (16.6)): graph σ, diagonal, and
the intersection, W. Thus the composite relation is essentially the figure, (16.6), with S "blown-down" to W. As we
pointed out in §15, σ is *continuous* as an involution of
T*P-0, but it is *not smooth* on S. Hence, on this blown—down
version of (16.6) the singularities are not just "normal
crossings." The figure below is a schematic depiction of the
composite relation in (T*P-0)x(T*P-0).

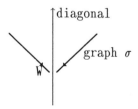

(The bent line is continuous but not C^1 at W.)

    To understand how to interpret these microlocal pictures,
we need to look at the problem of composing R_F and R_F^t from
a slightly more general perspective. Lets ask ourselves what
happens when we try to compose *any two integral transforms of
the type described in* §14: Suppose

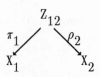

and

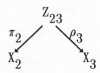

are double fibrations in the sense of §14, (and, to simplify a little the discussion below, lets assume that ρ_2 is *proper*). There is an obvious way of "composing" (16.7) and (16.8); namely we can regard Z_{12} and Z_{23} as relations between X_1 and X_2, and X_2 and X_3, and let Z_{13} be the composite relation;

(16.9) $Z_{13} = \{(x_1, x_3),\ \exists\ x_2$ such that

$$(x_1, x_2) \in Z_{12} \quad \text{and} \quad (x_2, x_3) \in Z_{23}\}$$

with the projection mappings

sending (x_1, x_3) to x_1 and x_3. Unfortunately, (16.10) will not in general be a double fibration. (In fact (16.9) will not, in general, be a *submanifold* of $X_1 \times X_3$.) However, the situation is a little better if we replace the set-theoretic composite relation, (16.9), by the "category theoretic" com-posite relation:

(16.12)

$$
\begin{array}{ccc}
 & W_{13} & \\
\pi \swarrow & & \searrow \rho \\
X_1 & & X_3
\end{array}
$$

sending (x_1, x_2, x_3) to x_1 and x_3. A useful way of viewing (16.11) is in terms of the mapping

(16.13) $\rho_2 \times \pi_2 \colon Z_{12} \times Z_{23} \rightarrow X_2 \times X_2 \ .$

The set (16.11) is just the pre-image, with respect to this mapping, of the diagonal in $X_2 \times X_2$. Since (16.13) is a fiber mapping, this description makes clear that (16.11) is a sub-manifold of $Z_{12} \times Z_{23}$. In fact it is clearly a submanifold of codimension equal to the dimension of X_2. Notice also that π and ρ, in figure (16.12), are fiber mappings, and that $\pi \times \rho$ maps W_{13} onto the set (16.9) in $X_1 \times X_2$. In general, however, this mapping will *not* be a differentiable imbedding of W_{13} into $X_1 \times X_3$. Therefore, though the diagram (16.12) has some properties of a double fibration, it fails in general to satisfy the "imbedding axiom." In particular, the points of

X_3 are indexing parameters for certain submanifolds of W_{13}, but these submanifolds are *not* mapped by π diffeomorphically into submanifolds of X_1. What is true, however, is that one can still associate integral transforms with the diagram (16.12); i.e. given a smooth measure, μ_{13}, on W_{13}, one can define an integral transform, $R: C_0^\infty(X_1) \to C^\infty(|\Lambda|X_3)$, by setting

$$(16.14) \qquad\qquad Rf = \rho_* \mu_{13} \pi^* f$$

just as in (14.10). We will show, in fact, that if R_{12} and R_{23} are integral transforms of the type (14.10) associated with the double fibrations (16.7) and (16.8), then their composition is an operator of the form (16.14). To prove this we will need the following elementary lemma, for the proof of which we refer to [15]:

Lemma 16.2. Let

$$
\begin{array}{ccc}
Z & \xrightarrow{\ \pi\ } & X \\
\rho \downarrow & & \downarrow g \\
Y & \xrightarrow{\ f\ } & W
\end{array}
$$

be a *transverse square*. (This means $f \times g$ is transverse to the diagonal in $W \times W$, and $\pi \times \rho$ imbeds Z into $X \times Y$ as the *pre-image* of the diagonal.) Let μ_1 and μ_2 be smooth densities on X and Y respectively and ν a smooth non-vanishing density on W. Then there exists a unique smooth density, κ, on Z such that for all functions, $\phi \in C_0^\infty(X)$

$$(16.15) \qquad f^*(\nu^{-1}(g_*\phi\mu_1))\mu_2 = \rho_*(\pi^*\phi\kappa) \ .$$

Now let μ_{12} be a smooth density on Z_{12} and μ_{23} a smooth density on Z_{23}, and let R_{12} and R_{23} be the integral operator

$$\phi \rightarrow (\rho_2)_*(\pi_1^*\phi\mu_{12})$$

and

$$\psi \rightarrow (\rho_3)_*(\pi_2^*\psi\mu_{23})$$

mapping $C_0^\infty(X_1)$ to $C_0^\infty(|\Lambda|X_2)$ and $C_0^\infty(X_2)$ to $C^\infty(|\Lambda|X_2)$ respectively. If ν is a smooth non-vanishing density on X_2, we can define the composition, R_{13}, of R_{12} with R_{23} setting

$$(16.16) \qquad R_{13}\phi = R_{23}\nu^{-1}R_{12}\phi \ .$$

We will prove:

Proposition 16.3. The operator (16.16) is an operator of the form (16.14); i.e. there exists a unique smooth measure, μ_{13}, on W_{13} with the property that

$$R_{23}\nu^{-1}R_{12}\phi = \rho_*\mu_{13}\pi^*\phi$$

for all $\phi \in C_0^\infty(X_1)$.

<u>Proof</u>. Apply lemma 16.2 to the transverse square

$$
\begin{array}{ccc}
W_{13} & \xrightarrow{\ \gamma\ } & Z_{12} \\
\tau \downarrow & & \downarrow \rho_2 \\
Z_{23} & \xrightarrow[\ \pi_2\ ]{} & X_2
\end{array}
$$

γ and τ being the mappings, $(x_1,x_2,x_3) \to (x_1,x_2)$, and $(x_1,x_2,x_3) \to (x_2,x_3)$, respectively. By lemma 16.2 there exists a smooth density, μ_{13}, on W_{13} with the property that for all smooth functions, $\phi \in C_0^\infty(Z_{12})$,

$$\pi_2(\nu^{-1}((\rho_2)_*\phi\mu_{12}))\mu_{23} = \tau_*((\gamma^*\phi)\mu_{13})\ .$$

Now substitute $\pi_1^*\phi$ for ϕ in this formula and apply the operator $(\rho_3)_*$ to both sides. Then the left hand side becomes

$$R_{23}\nu^{-1}R_{12}\phi$$

and the right hand side becomes $\rho_*(\pi^*\phi\mu_{13})$ in view of the fact that $\rho = \rho_3 \circ \tau$ and $\pi = \gamma \circ \pi_2$.   Q.E.D.

Lets now see what this composition formula says when we take the double fibrations (16.7) and (16.8) to be

(16.17)

$$\begin{array}{ccc} & Z & \\ {}^{\rho}\swarrow & & \searrow^{\pi} \\ P & & M \end{array}$$

and

(16.18)

$$\begin{array}{ccc} & Z & \\ {}^{\pi}\swarrow & & \searrow^{\rho} \\ M & & P \end{array}$$

In this case, W_{13} is the set

(16.19) $Y = \{(p,m,p'), \ m \in \gamma_p \cap \gamma_{p'}\}$,

γ_p and $\gamma_{p'}$ being the null-geodesics indexed by p and p'. Alternatively, let F_m be the submanifold of P indexed by m, using the double fibration (16.17). (i.e. F_m is the set of null—geodesics containing m.) Then

(16.20) $Y = \{(p,m,p'), \ p,p' \in F_m\}$.

Notice now that if $p \neq p'$, the null-geodesics, γ_p and $\gamma_{p'}$, are distinct; so they intersect in *exactly one* point. Thus, the mapping

$$j_0 \colon Y \to P \times P$$

sending (p,m,p') to (p,p') is an *imbedding* of Y into $P \times P$ away from the diagonal. On the other hand, if $p = p'$, the circle,

$$\{(p,m,p), \ m \in \gamma_p\}$$

gets "blown-down" by j_0 onto the diagonal point, (p,p). There is, however, a simple way of rectifying this problem: Let $(P \times P)_\Delta$ be the manifold obtained by blowing up $P \times P$ projectively along its diagonal and let

(16.21) $\beta \colon (P \times P)_\Delta \to P \times P$

be the blowing-up mapping. We will show that there is an honest imbedding

$$j \colon Y \to (P \times P)_\Delta$$

with $\beta \cdot j = j_0$. In fact its clear by continuity that if j exists it has to be unique. To show that it exists lets recall that, as an abstract set, $(P \times P)_\Delta$ consists of $P \times P - \Delta$ and the

singular locus, S_Δ of β; and since β is a projective
blowing-up, there is a canonical isomorphism of fiber spaces

(16.22)
$$S_\Delta \xleftarrow{\quad h \quad} \mathbb{P}(TP)$$
$$\pi_1 \searrow \quad \swarrow \pi_{\tan}$$
$$P$$

Now define the map, j, by setting

$$j(p,m,p') = \beta^{-1}(p,p') \qquad \text{if} \quad p \neq p'$$

(16.23)   and

$$j(p,m,p') = h(p,\ell) \qquad \text{if} \quad p = p'$$

where ℓ is the tangent line to F_m at p. It is easy to see
that j is smooth and imbeds Y into $(P \times P)_\Delta$ as a smooth
submanifold. We will henceforth think of Y as a submanifold
of $(P \times P)_\Delta$.

Now lets go back to the composite operator, (16.1). Let
ρ_1 and ρ_2 be the projections of Y onto P sending
(p,m,p') to p and p' respectively. By lemma 16.2 there
exists a measure, κ, on Y such that the operator (16.1) is
identical with the operator

(16.24) $$R_\kappa g = (\rho_2)_* (\kappa \rho_1^* g) \ .$$

We can think of κ, however, as defining a "delta–distribution" on $(P \times P)_\Delta$ with support on Y; i.e. we can define a distribution, δ_κ, on $(P \times P)_\Delta$ by means of the formula

$$\langle \delta_\kappa, f \rangle = \int_Y j^* f d\kappa \ .$$

The content of the formula (16.24) is:

<u>Proposition 16.3</u>. Let K be the Schwartz kernel of the operator (16.1), then

$$(16.25) \qquad\qquad K = \beta_* \delta_\kappa.$$

As a corollary of this proposition we obtain the following formula for the *pull-back* of K to $(P \times P)_\Delta$:

$$(16.26) \qquad\qquad \beta^* K = \beta^* \beta_* \delta_\kappa \ .^*$$

This formula tells us a lot about $\beta^* K$. In the first place, by composition formulae for wave-front sets, it tells us that the microlocal support of $\beta^* K$ consists of the three intersecting canonical relations depicted in the figure below:

---

*The definition of β^* is reviewed in the appendix.

the conormal bundle of Y, the conormal bundle of S_Δ, and the conormal bundle of the intersection, $Y_\Delta = S_\Delta \cap Y$

(16.27)

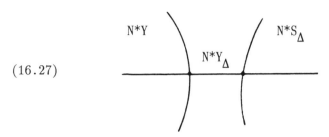

(Notice, by the way, that Y and S_Δ are hypersurfaces in $(P \times P)_\Delta$ and that they intersect transversally in Y_Δ. In particular, Y_Δ is a codimension two submanifold of $(P \times P)_\Delta$.) This, however, is not all that can be read off from (16.26). In [26], Melrose and Uhlmann have given a complete microlocal characterization of distributions of the form (16.26). They are examples of what Melrose and Uhlmann call *paired Lagrangian distributions*. These distributions have essentially all the properties of ordinary Lagrangian distributions, i.e. a nice symbolic calculus, functoriality with respect to push-forward and pull-back operations, reasonable behavior with respect to tensor products etc. (In particular Uhlmann has shown that the standard Sobolev estimates for F.I.O.'s are also true for F.I.O.'s with paired Lagrangian distributions as Schwartz kernels.) We won't have time to review this theory here,[*] but

---

[*] There is unfortunately not yet a convenient reference for the theory of paired Lagrangian distributions summarizing all the main facts. Some relevant papers are [16], [26], [42] and [43].

what we do want to stress is that by "blowing-up" K by pull-
ing it back to $(P \times P)_\Delta$, we obtain a distribution of a type
which is well-understood microlocally.

Let us next try to reconcile the micro-local picture,
(16.27), with the microlocal picture, (16.6). We will first
prove:

Proposition 16.4. There is a canonical conic diffeomorphism of
N*Y-0 in figure (16.27) onto graph σ in figure (16.6).

Proof. Let Y_0 be the complement of $Y \cap S_\Delta$ in Y. The
restriction of β to Y_0 is an imbedding of Y_0 into $P \times P - \Delta$,
and, by proposition 16.3, N^*Y_0 is the microlocal support of
K with its diagonal points deleted. On the other hand, by
proposition 16.1, this set is also the graph of σ with its
diagonal points deleted; so its clear that this proposition is
true if we stay away from the diagonal. We will next show that
this correspondence *extends smoothly over the diagonal*. To see
this notice first that the diagonal points in N*Y are the
points:

(16.28) $\{(Y,\xi),\ y \in Y_\Delta,\ \xi$ conormal to Y in $(P \times P)_\Delta$ at $y\}$.

Since Y and S_Δ intersect transversely in Y_Δ, (16.28) can
be identified with the set

(16.29) $\{(y,\xi),\ y \in Y_\Delta,\ \xi$ conormal to Y_Δ in S_Δ at $y\}$.

Now recall that P is a contact manifold: At each point, $p \in P$, a one-dimensional subspace, W_p, is singled out in the cotangent space, T_p^*. Moreover, by proposition 15.5, the F_m's are Legendrian submanifolds of P: if $p \in F_m$, then the tangent space to F_m at p is contained in the annihilator space, $W_p^\perp$, of W_p at p.

By identifying Δ with P we can think of S_Δ as a fiber bundle over P; indeed, by (16.22) there is an isomorphism of fiber bundles

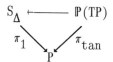

In particular the fiber, $\pi_2^{-1}(p)$, is just $\mathbb{P}(T_p)$. Therefore, by (16.23), the fiber of Y_Δ above p is $\mathbb{P}(W_p^\perp)$; and hence the set, (16.29), can be identified with the set

(16.30) $\{(y,(d\pi_1)_y^t\eta),\ \eta \in W_p$ and $p = \pi(y)\}$.

Now let ρ be the fibering of Z over P in (16.17). We showed in §15 that the subset, S, of Γ can be identified with

(16.31) $\{(z,(d\rho_z)^t\eta), \eta \in W_p \text{ and } p = \rho(z)\}$.

Define a mapping of (16.30) onto (16.31) by mapping
$(y,(d\pi_1)^t\eta)$ to $(z,(d\rho_z)^t\eta)$, where $y = (p,m,p)$ and $z =$
(p,m). We leave for the reader to check that this is a
diffeomorphism of (16.30) onto (16.31) and that it extends the
diffeomorphism defined above to the diagonal. Q.E.D.

The two other pieces of figure (16.27) are related to
figure (16.6) in a slightly more complicated way: The blowing–
up procedure gives rise to a correspondence between micro–
functions supported on $N^*S_\Delta-0$ in $T^*(P\times P)_\Delta$ and micro–
functions supported on $N^*\Delta-0$ in $T^*(P\times P)$; however, this
correspondence is "*micro*-microlocal," it is not a point-point
correspondence but is, itself, given by a Fourier integral
operator. Now if we identify $N^*\Delta$ with T^*P and T^*P-W with
$\Gamma-S$, micro-functions supported on $N^*S_\Delta-N^*Y_\Delta$ in figure (16.27)
are in one-one correspondence with micro-functions supported on
$\Gamma-S$ in figure (16.6). (All this will be spelled out a lot
more carefully in the appendix.) Finally the "bridge" between
N^*Y and N^*S_Δ in figure (16.27), i.e. the set N^*Y_Δ, gets
sent by blowing-down into the conormal bundle of W in T^*P.
(Again this will be spelled out more carefully in the
appendix.)

<u>A final remark</u>. These results extend in a straightforward way to integral transforms involving forms or sections of vector bundles. They are also true if we replace integral transforms by micro-localized integral transforms, e.g. F.I.O.'s defined on Γ, Γ^t etc.

There are two appendices to this section. In appendix B, we discuss some of the micro-local aspects of "blowing-up"; and, in appendix A, we show that, for the standard $M_{2,1}$. $R_\mu \circ R_\mu^t$ is much better behaved than for general M and that its main properties can be easily described without recourse to microlocal techniques.

__Appendix A__. Let $M = M_{2,1}$ = standard compactified Minkowski space and let

(A16.1)

$$\rho \swarrow \overset{Z}{} \searrow \pi$$
$$P \qquad M$$

be the double fibration associated with the standard x-ray transform. Let $G = U(2)$ = the maximal compact subgroup of $Sp(2,\mathbb{R})$. Since G acts transitively on Z, M and P, there exist unique G-invariant probability measures, μ, ν_1 and ν_2, on these spaces. Consider the integral operator

(A16.2) $RR^t \colon C^\infty(P) \to C^\infty(P)$

defined by the formula:

(A16.3) $RR^t f = \nu_2^{-1} \rho_* \mu \pi^* \nu_1^{-1} \pi_* \mu \rho^* f$.

In this section we will discuss the basic properties of this operator from a "classical" (i.e. non-microlocal) point of view. To begin with we will examine the double fibration

(A16.4)

$$\pi_1 \swarrow \overset{W_0}{} \searrow \pi_2$$
$$P \qquad P$$

obtained by composing (A16.1) with its transpose. By (16.11)

(A16.5) $W_0 = \{(p,m,q), (p,q) \in F_m\}$,

π_1 and π_2 being the projections, $\pi_1(p,m,q) = p$ and $\pi_2(p,m,q) = q$. By proposition 16.3, there is a G-invariant measure, κ, on W, canonically associated with μ and ν_1, with the property that (A16.2) is given by the formula,

(A16.6) $RR^t f = \nu_2^{-1}(\pi_1)_* \kappa (\pi_2)^* f$.

To obtain a more concrete picture of this transform, lets describe the fibration (A16.4) from the point of view of the "Kleinian picture of $M_{2,1}$" presented in §2 and §9. In this picture, $M_{2,1}$ is the Grassmannian of Lagrangian subspaces of $\mathbb{R}^4$, P is $\mathbb{R}P^3$ and the F_m's are just the projective lines in $\mathbb{R}P^3$ associated with Lagrangian subspaces of $\mathbb{R}^4$, in other words, "Legendrian" projective lines in $\mathbb{R}P^3$. Now consider the map

$$\pi_1 \times \pi_2 : W_0 \to P \times P \ .$$

Let W be the image of this map. If $(p,q) \in W$, there has to exist a Legendrian line, F_m, containing both p and q; hence the set

(A16.7) $W_p = \{q, (p,q) \in W\}$

is just the union of all Legendrian lines emanating from p.
Notice that this set is a 2-dimensional projective subspace of
$\mathbb{RP}^3$. In fact, suppose p is the point in projective spaces
corresponding to the one-dimensional vector subspaces, L, of
$\mathbb{R}^4$. Then

(A16.8) $W_p = Proj(L^\perp)$,

$L^\perp$ being the symplectic orthocomplement of L in $\mathbb{R}^4$, and W_p
its projective image in $\mathbb{RP}^4$. Now consider the fiber $(W_0)_p =$
$\pi^{-1}(p)$ in the diagram (A16.4) above p. By definition $(W_0)_p$
consists of all pairs (ℓ, q) where ℓ is a Legendrian line
containing p and q a point on this line. The mapping

(A16.9) $\pi_2 \colon (W_0)_p \to W_p$

sending (ℓ, q) to q is 1-1 except on the circle

$$\{(p, \ell), \ p \in \ell\}$$

and it is easy to see that (A16.9) is, in fact, a projective
"blowing up" of W_p at p.

Given the measure, κ, on W_1 and the density, (ν_2), on T_p, one can canonically associate with these two pieces of data a smooth non-vanishing measure, κ_p^1, on the fiber of π_1 above p (i.e. on $(W_0)_p$) so that, roughly speaking, $(\kappa_p^1)_q$ and $(d\pi_1)_q^t(\nu_2)_p$ are the vertical and horizontal parts of κ at q for any point, q, on this fiber. Let κ_p be the push-forward of κ_p^1 with respect to the blowing-up map (A16.9). This measure is an absolutely continuous measure on W_p, but it is no longer smooth. In fact we showed in the appendix to §14 that it has to have the form

$$(16.10) \qquad\qquad \kappa_p = \frac{1}{r}(r dr d\theta)$$

where r and θ are G_p invariant polar coordinates on W_p centered at p.

Coming back to the integral transform, (A16.2) we can, in view (A16.6) and (A16.9), write it more concretely in the form

$$(A16.11) \qquad\qquad (RR^t f)(p) = \int_{W_p} f d\kappa_p$$

In this form it resembles the classical Radon transform on $\mathbb{RP}^3$ (see Helgason, [19]) except for the fact that κ_p, instead of being the standard smooth spherical measure on W_p, is the singular measure (A16.10).

Before discussing its properties, it is useful to note that it belongs to a family of integral transforms

$$(A16.12) \qquad T_\alpha f(p) = \int_{W_p} f d\kappa_p^\alpha$$

associated with the measures

$$(A16.13) \qquad \kappa_p^\alpha = r^\alpha(rdrd\theta) \ ,$$

and that some of these other transforms are themselves of considerable interest. The following three examples are particularly important:

Example 1, $\alpha = 1$. Consider the integral transform (14.12)

$$R: \ \Omega^1(M) \rightarrow C^\infty(P)$$

defined by

$$(R\omega)(p) = \int_{\gamma_p} \omega$$

Since M isn't oriented, the intrinsic transpose of R is a transformation from $\Omega^3(P)$ into $\Omega^2(M) \otimes \theta$ where θ is the orientation bundle of M. (In DeRham's language R^t maps 3-forms on P into "impaire" 2-forms on M.) Now equip M

with the unique G invariant Lorentz metric compatible with
its canonical (Sp(2,ℝ)-invariant) causal structure, and let

$$*: \ \Omega^2(M) \ \otimes \ \theta \ \to \ \Omega^1(M)$$

be the Hodge star operator associated with this metric. We
will see in §17 that the operator, $R*R^t$, is just the operator,
T_α for $\alpha = 1$.

Example 2, $\alpha = -2$. As it stands T_α isn't defined for $\alpha = -2$
since r^{-2} isn't integrable. However, one can still define
T_{-2} by interpreting the integral, (A16.12), in an appropriate
generalized sense. This operator is the standard example of a
Radon singular integral operator. These operators are studied
in detail by Phong and Stein in [33]. Among other things they
show that if one considers P as the boundary of the unit ball
in $\mathbb{C}^2$ (with antipodal points identified) there is an
interesting factorization of the Green's operator of the $\bar{\partial}_b$-
Laplacian in which T_{-2} is one factor. Their result can be
interpreted, incidentally, as an inversion formula for T_{-2}.

Example 3, $\alpha = 0$. We've already commented on this example. T_0
is the standard Radon transform on $\mathbb{RP}^3$. An inversion formula
for it can be found in [19].

Since the T_α's are G invariant, one can study their
properties by decomposing $C^\infty(P)$ into G irreducible subspaces. On each of these subspaces T_α is equal to the
identity times a scalar multiplier, and hence T_α is completely determined by these multipliers. For $\alpha = -2$, 0 and 1,
i.e. for the three examples above, these multipliers are all
non-zero (which explains why there are inversion formulas in
these cases). However, for $\alpha = -1$, i.e. for the operator,
(A16.2), roughly half of these multipliers are zero; so (A16.2)
fails (rather spectacularly) to be invertible. This fact was
pointed out several years ago by Gelfand and Graev in a
slightly different context from ours. Namely they showed that
the non-invertibility of (A16.2) has to do with the fact that
the general line complex of order one in $\mathbb{RP}^3$ is not admis-
sible. (See [11].) Another explanation for the anomalous
behavior of $\alpha = -1$ is the following: One can show that for
$\alpha > -1$, T_α behaves essentially like the classical Radon
transform, i.e. doesn't "see" the singularities of κ_p^α at p.
On the other hand for $\alpha < -1$ the singularities of κ_p^α are
sufficiently sharp that they force T_α to behave as if κ_p^α
were an atomic measure concentrated at p; i.e. behave like a
pseudodifferential operator. Thus at $\alpha = -1$, T_α changes
from being a pseudodifferential operator to being a Radon
integral operator!

Appendix B (Microlocal aspects of blowing-up). We will begin with the following simple example: the blowing-down map

$$f: \mathbb{R}^2 \to \mathbb{R}^2$$

defined by $f(r,s) = (x,y)$ where

(B16.1) $x = r, \quad y = rs.$

(Notice that the line, $r = 0$, gets blown-down by f to the origin in (x,y) space.) Let ξ, η, ρ and σ be the cotangent coordinates dual to x, y, r and s. The canonical relation associated with f is given in these coordinates by

$$x = r, \qquad y = rs$$
(B16.2)   and
$$\rho = \xi + \eta s, \quad \sigma = \eta r \ .$$

From these equations one can easily read off the following properties of this "microlocalized" version of f:

    1. There are points on this canonical relation of the form: $(\xi,\eta) \neq 0$ and $(\rho,\sigma) = 0$. Indeed setting $\rho = 0$ and $\sigma = 0$ in (B16.2), one sees that the set of all such points is given by the equations

(B16.3) $\rho = \sigma = x = y = r = 0$

        and

(B16.4) $(\xi, \eta) = \lambda(s, -1), \quad \lambda \neq 0.$

The existence of such points means that if we "blow down" elements of $C_0^\infty(\mathbb{R}^2)$ by the push–forward operation, f_*, the objects we get are not C^∞ at $(0,0)$. It also means that the pull–back operation, f^*, is not well–defined on $C^{-\infty}(\mathbb{R}^2)$. To make it well-defined one has either to use microlocal cut-offs or regard the pull-back, $f^*\phi$, of a distribution, ϕ, as being a microfunction on the *complement* of the set, (B16.3)-(B16.4).

2. The relation, (B16.2), is defined by the diagram

(B16.5)

where Γ_f is (r, s, ξ, η)-space, and F and G are the maps:

$$F(r, s, \xi, \eta) = (x, y, \xi, \eta)$$

and

$$G(r, s, \xi, \eta) = (r, s, \rho, \sigma) \ .$$

It is clear from (B16.2) that F and G are themselves
"blowing-down" maps. In fact G blows the set, r = 0, down
onto the conormal bundle, $\sigma = 0$, r = 0, of the line, r = 0, in
(r,s)-space; and F blows the set, r = 0, down onto the cotangent space to the origin in (x,y)-space. Off the set,
r = 0, Γ is the graph of the canonical transformation

$$(r,s,\rho,\sigma) \rightarrow (x,y,\xi,\eta)$$

defined by the equations:

(B16.6) $x = r, \quad y = rs, \quad \xi = \rho - \sigma(s/r), \quad \eta = \sigma/r$.

This canonical transformation has the appearance of being
highly discontinuous at r = 0; however, we will see below that
there is a "micro-microlocal" mechanism built into the diagram
(B16.5) which controls these discontinuities.

3. In addition to the points (B16.2) one has to include
as part of the canonical relation, Γ_f, certain limit points.
Explicitly, let p_0 be the point, $x = y = \xi = 0$, $\eta = 1$, in
(x,y,ξ,η) space, (i.e. the conormal vector to the line y = 0
at the origin), and let p(t) be the point, $x = bt$, $y = abt^2$,
$\xi = 0$, $\eta = 1$. As t tends to zero p(t) tends to p_0 along
a curve which is tangent at p_0 to the space $\xi = y = 0$ (i.e.

the conormal bundle of the line $y = 0$). But by (B16.2) the
image of this curve in (r,s,ρ,σ) space is

$$(r,s) = t(b,a) \quad \text{and} \quad (\rho,\sigma) = t(a,b) \ .$$

Therefore, it is clear that one has to include as limit points
of (B16.2) the set

(B16.7) $x = y = \xi = r = s = 0, \qquad \rho,\sigma,\eta$ arbitrary.

In particular, suppose that μ is a "paired Lagrangian" dis-
tribution associated with the two Lagrangian manifolds, $\Lambda_0 = T_0^*$ and $\Lambda_1 = N^*(y = 0)$, in (x,y,ξ,η) space. (For instance,
$\mu =$ the fundamental solution, $H_+(x) \otimes \delta(y)$, of the
differential equation, $\frac{\partial}{\partial x} \mu = \delta$.) Then the microlocal support
of μ in (x,y,ξ,η) space is given by the figure:

$\Lambda_0 \qquad \Lambda_1$

but its pre-image in (r,s,ρ,σ) space has the microlocal
support shown in the figure:

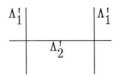

where $\Lambda_1' = N^*(r = 0)$, $\Lambda_2' = N^*(r = s = 0)$ and $\Lambda_1' = N^*(s = 0)$. (compare with the figures (16.6) and (16.27) in the text.) Λ_1' and Λ_1 are now disjoint, but the limit set forms a "bridge," Λ_2', joining Λ_1' to Λ_1'.

We will now give a slightly more systematic account of these matters. Let X and Y be n-dimensional manifolds, S a hypersurface in X, W a codimension k submanifold of Y and

$$f: X \to Y$$

a smooth map which is a diffeomorphism off S and blows S down to W. The microlocal diagram associated with f is

(B16.8)
$$\begin{array}{ccc} & \Gamma_f & \\ F \swarrow & & \searrow G \\ T^*Y & & T^*X \end{array}$$

where $\Gamma_f = f^*T^*Y$ and F and G are the maps:

$$F(x,\eta) = (f(x),\eta)$$

and

$$G(x,\eta) = (x,(df_x)^t\eta) \ .$$

Notice that $(x,0,y,\eta)$ is a point on this canonical relation if and only if

(B16.9) $x \in S$, $y = f(x) \in W$ and $\eta \in \ker(df_x)^t$.

In other words one has to delete this set from Γ in order for the operations f_* and f^* to be well-defined. Let Γ_0 be the hypersurface

$$\Gamma_0 = f^*T^*Y \,\big]\, S$$

in Γ and let Σ_1 and Σ_2 be the codimension k submanifolds

$$\Sigma_1 = T^*Y \,\big]\, W$$

and

$$\Sigma_2 = \{(x,\xi) \in T^*X, \ \xi \perp \ker df_x\}$$

of T^*Y and T^*X respectively. Then, away from Γ_0, both F and G are diffeomorphisms, and hence Γ is the graph of a canonical transformation. On the other hand, F maps Γ_0

onto Σ_1 and G maps Γ_0 onto Σ_2, and it is easy to show that both these maps are "blowing-down" mappings. Notice also that Σ_1 and Σ_2 are co-isotropic, and the "reduced" sym-plectic manifold which one obtains by identifying leaves of the null-foliation is, in both cases, identical with T*W i.e. one has the reduction diagram

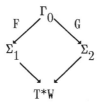

To get a better feeling for what is involved in the microlocal version, (B16.8), of f, it is useful to forget for the moment that T*X and T*Y are cotangent bundles and consider (B16.8) as a special case of the following set-up: a pair of sym-plectic manifolds, M and N, of dimension 2n, a pair of co-isotropics, Σ_1 in M and Σ_2 in N, both of codimension, $k \leq n$, a canonical relation,

and a hypersurface, Y, in Γ such that π and ρ are maps which blow Y down to Σ_1 and Σ_2 respectively. On the

complement of Y, π and ρ have to be diffeomorphisms
locally, and for simplicity, we will assume that Γ-Y is the
graph of a canonical transformation

(B16.11) Φ: M-Σ_1 $\rightarrow$ N-Σ_2 .

Let NY be the normal bundle to Y in Γ and $N\Sigma_1$ and $N\Sigma_2$
the normal bundles to Σ_1 in M and Σ_2 in N respectively.
One feature of "blowing-down" is that there are natural maps

$$d\pi\colon NY\text{-}0 \rightarrow N\Sigma_1\text{-}0$$

and

$$d\rho\colon NY\text{-}0 \rightarrow N\Sigma_2\text{-}0$$

which are, locally, diffeomorphisms. Let us make a further
simplification by assuming that $d\pi$ and $d\rho$ are *global*
diffeomorphisms. This enables us to define a composite
diffeomorphism

$$\Psi\colon N\Sigma_1\text{-}0 \rightarrow N\Sigma_2\text{-}0 \ ,$$

the composite of $(d\pi)^{-1}$ and $d\rho$. The map, Φ, is discontinuous at Σ_1 (as we noticed for instance in the example
(B16.6)); however, its discontinuities are controlled by Ψ in
the sense that if $\{p_i\}$ is a sequence of points in M

approaching the point $p_0 \in \Sigma_1$ along the normal direction, n_0, the image sequence, $\{\Phi(p_i)\}$, approaches $p' \in \Sigma_2$ along the normal direction, n', where $(p',n') = \Psi(p_0,n_0)$.

We will show that the map Ψ is a "micro-microlocal" feature of the diagram (B16.10). More explicitly, suppose the co—isotropics, Σ_1 and Σ_2, are *Lagrangian* submanifolds of M and N. (i.e. Consider the symplectic analogue of "blowing-up" a point.) Then there are canonical symplectic identifications

$$N\Sigma_1 \cong T^*\Sigma_1$$

and

$$N\Sigma_2 \cong T^*\Sigma_2 \ ,$$

and hence Ψ can be regarded as a mapping

(B16.12) $\Psi: T^*\Sigma_1\text{-}0 \rightarrow T^*\Sigma_2\text{-}0$

We claim that this mapping is a *canonical transformation*. In fact it is an easy exercise, which we leave for the reader, to show that (B16.12) is the canonical transformation associated with the double fibration

(We saw in section 14, that, given a double fibration of this type, one gets a canonical relation of the type (14.6). In the situation above this canonical relation is a canonical graph, and, in fact, is simply the graph of (B16.12).)

Another important micro—microlocal feature of (B16.10) is the following: Let Λ_1 be a Lagrangian submanifold of M which intersects Γ cleanly. It follows from general facts about clean intersection ([20], Chapter XXV) that

$$\Lambda_2 = \{\rho(\gamma), \ \gamma \in \Gamma, \ \pi(\gamma) \in \Lambda_1\}$$

is an (immersed) Lagrangian submanifold of N. For instance, if Λ_1 is disjoint from Σ_1, then Λ_2 is the image of Λ with respect to the canonical transformation, Φ; and, at the other extreme, if $\Lambda_1 = \Sigma_1$ then $\Lambda_2 = \Sigma_2$. Now suppose Γ is equipped with a "symbol," i.e. a half-density, a half-form, or the tensor product of a half-density with a Maslov factor. Then (loc.cit.) there is a symbolic correspondence between symbols ($\frac{1}{2}$ densities, etc.) on Λ_1 and symbols on Λ_2. If Λ_1 is disjoint from Σ_1, this is just the point-point correspondence associated with Φ: if σ is, for instance, a

$\frac{1}{2}$-density on Λ_1 the corresponding $\frac{1}{2}$ density on Λ_2 is $(\Phi^{-1})^*\sigma$. On the other hand, if $\Lambda_1 = \Sigma_1$, this correspondence is much more subtle. One can show that if Γ is equipped with a $\frac{1}{2}$-density, then Y is equipped with a triple, (μ_1, μ_2, μ), where μ_1 is a $(-\frac{1}{2})$-density on Σ_1, μ_2 a $(-\frac{1}{2})$-density on Σ_2 and μ a density on Y. The correspondence between symbols on Σ_1 and symbols on Σ_2 is given by the integral transform

$$\sigma \to \mu_2 \rho_* \mu \pi^* (\mu_1 \sigma)$$

which is, as we know from §14, *a Fourier integral operator with* (B16.12) *as its underlying canonical transformation.* In other words this correspondence is *micro-microlocal!*

We will conclude with a few words about how these results generalize to co-isotropics. Suppose that the null-foliations of Σ_1 and Σ_2 are fibrating. One can show that if F_1 is a null-leaf of Σ_1 the set

$$F_2 = \{\rho(\gamma), \ \gamma \in \Gamma, \ \pi(\gamma) \in F_1\}$$

is a null-leaf of Σ_2. Thus, if X_1 is the symplectic reductions of Σ_1 by its (fibrating) null-foliation and X_2 the symplectic reduction of Σ_2, there is a map

$$h: X_1 \to X_2$$

One can show that this is (locally) a symplectomorphism. Let
us for simplicity assume that it is a global symplectomorphism
and use it to identify X_1 with X_1. We then have the
diagram:

(B16.13)

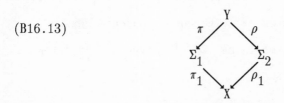

X being the common symplectic reduction of Σ_1 and Σ_2. Now
the normal bundle of Σ_1 in M has a canonical symplectic
identification with the cotangent bundle, $T^*\pi_1$, of the fibration, π_1, i.e. if $p \in \Sigma_1$ the normal space to Σ_1 at p is
identified by the symplectic form at p with the cotangent
space at p to the fiber through p of π_1. Similarly the
normal bundle of Σ_2 can be identified with $T^*\rho_1$ and hence
the map $\Psi: N\Sigma_1-0 \rightarrow N\Sigma_2-0$ gives rise to a map

(B16.14) $\Psi: T^*\pi_1-0 \rightarrow T^*\rho_1-0$.

Now let $q \in X$ and let F_1 and F_2 be the fibers of π_1 and
ρ_1 above q. Since (B16.13) commutes, F_1 and F_2 have a
common pre-image, Z, in Y, and by restricting π and ρ to
fibers above q we get a double fibration

(B16.15)

On the other hand (B16.14) restricts to a mapping

$$\Psi: T^*F_1-0 \rightarrow T^*F_2-0 \ .$$

One can prove that Ψ is a canonical transformation and that, in fact, just as in the Lagrangian case, it is the canonical transformation associated with the double fibration, (B16.15).

§17   In this section we will compute the symbols of the operators, R_F and R_F^t, and describe how to construct parametrices for these operators. We will make heavy use in the next few pages of Hörmander's half-density notation and the symbolic calculus developed in Chapter XXV of his book. Apropos of this, in §14 we used the symbol, Ω^k, which is Hörmander's standard symbol for "k-densities" (with $k = 1/2$ or 1), to denote "spaces of k-forms" (with $k = 0,1,2,\ldots$). We will use in its place the symbol, $|\Lambda|^k$; i.e. if M is a manifold, $|\Lambda|^k M$, will be, by definition, *the bundle of k-densities on* M, and $|\Lambda|_m^k$ its fiber at m (the one-dimensional space of k densities on T_m). We will also be a little cavalier about Maslov factors in the computations below.[*] As far as R_F and R_F^t are concerned, this is excusable: Since their distributional kernels are conormal distributions, the Maslov contributions to their symbols can be normalized to be equal to 1. The notation in §17 (M, P, Z, π, ρ, Γ, S, Σ etc.) is the same as in §§15 and 16. Finally, we will frequently have recourse below to the following elementary identification:

Lemma 17.1. At a point $g = (m,\gamma,p,\eta) \in \Gamma$, the space of half-densities, $|\Lambda|_g^{\frac{1}{2}}$, is canonically isomorphic to

---

[*]i.e. we will ignore them entirely.

(17.1)
$$|\Lambda|_p^{\frac{1}{2}} \otimes |\Lambda|_m^{-\frac{1}{2}} \otimes |\Lambda|_m(\gamma_p) \ ,$$

γ_p being the null-geodesic through m indexed by p.

<u>Proof</u>. Since Γ is the conormal bundle of Z in $M \times P$

(17.2)
$$|\Lambda|_g^{\frac{1}{2}} \cong |\Lambda|_Z^{\frac{1}{2}} \otimes |\Lambda|^{-\frac{1}{2}}(N_Z) \ ,$$

where N_Z is the normal space to Z at z in $M \times P$. However, by (14.8)

(17.3)
$$|\Lambda|^{\frac{1}{2}}(N_Z) \cong |\Lambda|_m^{\frac{1}{2}} \otimes |\Lambda|_m^{-\frac{1}{2}}(\gamma_p) \ .$$

Moreover, since γ_p is the fiber of the map, $\rho: Z \to P$, above p,

$$|\Lambda|_Z^{\frac{1}{2}} \cong |\Lambda|_p^{\frac{1}{2}} \otimes |\Lambda|_m^{\frac{1}{2}}(\gamma_p) \ .$$

Inserting this and (17.3) into (17.2), we get (17.1) as claimed.   Q.E.D.

We showed in §15 that the x-ray transform

$$R_F: \ C^\infty(T^*M) \to C^\infty(P)$$

from the space of one-forms on M to the space of functions on

P is a Fourier integral operator with the microlocal diagram

$$(17.4) \qquad\qquad \begin{array}{c} \Gamma \\ \pi \diagup \quad \diagdown \rho \\ T^*M\text{-}0 \quad T^*P\text{-}0 \end{array}$$

as its underlying canonical relation. Therefore, ([20], Ch.

XXV, Theorem 25.2.2) the same is true of the transpose operator

$$R_F^t\colon C^\infty(|\Lambda|(P)) \to C^\infty(TM \otimes |\Lambda|)$$

Let us fix, once and for all, positive densities, μ, on P

and, ν, on M and consider the operator

$$(17.5) \qquad \nu^{-\frac{1}{2}} R_F^t \mu^{\frac{1}{2}}\colon C^\infty(|\Lambda|^{\frac{1}{2}}(P)) \to C^\infty(TM \otimes |\Lambda|^{\frac{1}{2}}) \ .$$

By the conventions of [20], Ch. XXV, the symbol of this

operator at $g = (m,\xi,p,\eta) \in \Gamma$ is an element of

$$\mathrm{Hom}(\mathbb{R},T_m) \otimes |\Lambda|_g^{\frac{1}{2}} \ .$$

However, $\mathrm{Hom}(\mathbb{R},T_m) \cong T_m$, and $|\Lambda|_g^{\frac{1}{2}}$ is isomorphic to the pro-

duct, (17.1); so the symbol of (17.4) at g is an element of

$$(17.6) \qquad T_m \otimes |\Lambda|_p^{\frac{1}{2}} \otimes |\Lambda|_m^{-\frac{1}{2}} \otimes |\Lambda|_m(\gamma_p) \ .$$

Now let w be a vector tangent to γ_p at m and pointing in
the direction of positive orientation. Let κ_w be the unique
element of $|\Lambda|_m(\gamma_p)$ satisfying $\kappa_w(w) = 1$. (thus

$$(17.7) \qquad\qquad \kappa_{\lambda w} = \frac{1}{\lambda}\kappa_w$$

if $\lambda \in \mathbb{R}^+$.)

The following is left as an easy exercise for the reader:

<u>Lemma 17.2</u>. The symbol of the operator (17.5) at $g \in \Gamma$ is

$$(17.8) \qquad\qquad F(z)w \otimes \mu_p^{\frac{1}{2}} \otimes \nu_m^{-\frac{1}{2}} \otimes \kappa_w \ .$$

Notice that, by (17.7), this definition is independent of the
choice of w.

The construction of a parametrix for (17.5) will involve
a certain amount of microlocal tinkering with the operator R_F.
For this purpose we will introduce some elementary geometric
machinery on Z. Recall that, by defining a causal structure
on M, we have assigned to every point, m, a cone, C_m, in the
tangent space to M at m. This cone is *not* a Lorentzian
cone, i.e. it is not defined by a quadratic equation. However,
if r is a null—ray in T_m there is a unique Lorentzian cone,
C_r, which osculates to second order with C_m along r. For

each r let us choose an axis of symmetry for C_r as in the
figure below

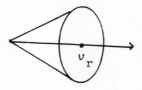

and let v_r be a point on this axis pointing in a positive
time-like direction. Let $\langle \cdot, \cdot \rangle_r$ be the Lorentzian inner
product on T_m compatible with C_r, normalized so that v_r
has unit length.

In §15 we showed that the fiber of Z over m is
canonically diffeomorphic to the set of null-rays on C_m. If
z is a point on this fiber we will write C_z for C_r, v_z
for v_r and $\langle \cdot, \cdot \rangle_z$ for $\langle \cdot, \cdot \rangle_r$. It is clear that one can
choose v_z and $\langle \cdot, \cdot \rangle_z$ to depend smoothly on z. (There is,
in fact, a slick way to do this: Equip M with a *Riemannian*
metric, $(\cdot, \cdot)$, and choose v_z to be the *true* axis of symmetry
of C_z with respect to $(\cdot, \cdot)_m$, i.e. let (e_1, e_2, e_3) be a
basis of T_m in which $\langle \cdot, \cdot \rangle_z$ and $(\cdot, \cdot)_m$ are simultaneously
in diagonal form and $\langle e_1, e_1 \rangle_z = -1$, and let $v_z = e_1$.)

We also showed in §15 that the fiber of Γ over m is
canonically diffeomorphic with the set

$$\Gamma_m = \{(r, \xi), \text{ r a null-ray on } C_m, \xi \in T_m^*, r \perp \xi\}$$

Moreover, (proposition 15.3) the involution, σ, maps Γ_m onto Γ_m sending (r,ξ) to (r',ξ) where r' is the unique null-ray, other than r, perpendicular to ξ. (The fixed point set of σ, under this identification, becomes the set of (r,ξ) for which ξ is itself a null-covector, in which case $r = r'$.)

   We will now introduce a function

$$H: \Gamma \to \mathbb{R}$$

which will play an important role in the symbolic computations below. Roughly speaking $H(r,\xi)$ will be the Lorentzian angle between r and r' measured in the $\langle \cdot , \cdot \rangle_{r'}$ frame. More explicitly

$$(17.9) \qquad\qquad H(r,\xi) = \langle w,w' \rangle_{r'}$$

where w and w' are vectors on the null-rays, r and r', and are normalized by requiring that $\langle w,v \rangle = \langle w',v' \rangle = -1$, where $v = v_r$ and $v' = v_{r'}$.

Lemma 17.3. H is strictly positive on Γ-S and vanishes exactly to second order on S.

Proof. Let L be a defining function for the convex cone C_m in T_m; i.e. let L be a function on T_m-0 which is homogeneous of degree 2 and defines C_m in the sense that C_m is one "nappe" of the set

$$\{v \in T_m, \ L(v) > 0\} \ .$$

Since C_m is strictly convex we can assume that

$$\det\left[\frac{\partial^2 L}{\partial y_i \partial y_i}\right] \neq 0$$

everywhere. By Euler's identity

(17.10) $$\sum \frac{\partial^2 L}{\partial y_i \partial y_j}(y) y_i w_j = \sum \frac{\partial L}{\partial y_j}(y) w_j \ .$$

If $y \in \partial C_m$ the tangent plane to the boundary along the ray through y is defined by setting the right hand side of (17.10) equal to zero. However, C_m itself lies entirely on one side of this plane; so if $w \in \partial C_m$, (17.10) is greater than or equal to zero with equality if and only if y and w are collinear. Morover, strict convexity requires that (17.10) have a strict minimum on the ray through y as one varies w. Q.E.D.

Now lets come back to the operator, (17.5), and describe how to construct a parametrix for it. Our first step will be to actually construct *two* parametrices for (17.5) one of which we will call the *on-diagonal parametrix* and the other the *off-diagonal parametrix*. We will begin with the off-diagonal parametrix which will, by definition, be an operator

$$(17.11) \qquad Q_\sigma \colon \; C^\infty(TM) \to C^\infty(P) \; ,$$

from vector fields on M to functions on P. Given a section, s, of TM let π^*s be its pull-back to Z. Let z be a point of Z and $m = \pi(z)$. The Lorentzian inner product, $\langle \cdot, \cdot \rangle_z$ on T_m gives one an identification

$$*_z \colon \; T_m \to T_m^* \; .$$

By applying $*_z$ to $(\pi^*s)(z)$ we convert $(\pi^*s)(z)$ into an element of T_m^* and by applying $(d\pi_z)^t$ to it we convert it into an element

$$(17.12) \qquad (d\pi_z)^t *_z (\pi^*s)(z)$$

of T_z^*. Its clear that (17.12) depends smoothly on z; that is, there exists a globally defined one-form

$$(d\pi)^t * (\pi^*s)$$

on Z whose value at z is (17.12). Now define $\mathbb{Q}_\sigma s$ by defining it at p to be the integral

$$(17.13) \qquad (\mathbb{Q}_\sigma s)(p) = \int_{\gamma_p} (d\pi)^t *(\pi^* s)$$

over the null geodesic γ_p.

To compose $\mathbb{Q}_\sigma$ with the operator (17.5) we first have to multiply it, fore and aft, by the half–densities, $\mu^{\frac{1}{2}}$ and $\nu^{-\frac{1}{2}}$; i.e. replace it by

$$(17.14) \qquad \mu^{\frac{1}{2}}\mathbb{Q}_\sigma \nu^{-\frac{1}{2}} : C^\infty(TM \otimes |\Lambda|^{\frac{1}{2}}) \to C^\infty(|\Lambda|^{\frac{1}{2}}P) \ .$$

Lets compute the symbol of (17.15): By definition this symbol is a function on Γ whose value at $g = (m,\xi,p,\eta)$ is an element of the space

$$(17.15) \qquad \text{Hom}(T_m,\mathbb{R}) \otimes |\Lambda|^{\frac{1}{2}}_g \ .$$

But, by lemma 17.1, this space is isomorphic to the space

$$(17.16) \qquad T_m^* \otimes |\Lambda|^{\frac{1}{2}}_p \otimes |\Lambda|^{-\frac{1}{2}}_m \otimes |\Lambda|_m(\gamma_p) \ ,$$

and it is not hard to see that, in terms of (17.16), the symbol of the operator (17.14) at $g \in \Gamma$ is just:

(17.17) $$(*_z)w \otimes \mu_p^{\frac{1}{2}} \otimes \nu_p^{-\frac{1}{2}} \otimes \kappa_w \; ,$$

where w is any non-zero vector tangent to γ_p at m point-ing in the direction of positive orientation, and κ_w is the unique element of $|\Lambda|_m(\gamma_p)$ satisfying $\kappa_w(w) = 1$. (Compare with 17.8.)

To define the on-diagonal parametrix, we will reverse the sequence of events above: we will first define its *symbol*, and then define the operator itself to be any operator

(17.18) $$Q_\Delta : \; C^\infty(TM) \to C^\infty(P)$$

having this symbol. In analogy with (17.17) we will define $\sigma(Q_\Delta)$ at $g = (m, \xi, p, \eta)$ to be the expression,

(17.19) $$*_{z'}(w') \otimes \kappa_w$$

the notation here being the following: Let $\sigma(m, \xi, p, \eta) = (m, \xi, p', \eta')$. Then $z' = (m, p')$; w and κ_w are the same as in (17.17), and w' is a vector tangent to $\gamma_{p'}$ at m whose projection onto the cone axis in $C_{z'}$ has the same length as the projection of w onto the cone axis in C_z. Notice that if we define Q_Δ by (17.17), then the symbol of $\mu^{\frac{1}{2}}Q_\Delta\nu^{-\frac{1}{2}}$ at g is, except for the primes, identical with (17.17):

$$(17.20) \qquad \sigma(\mu^{\frac{1}{2}}Q_\Delta \nu^{-\frac{1}{2}})_g = *_{z'}(w') \otimes \nu_p^{\frac{1}{2}} \otimes \nu_m^{-\frac{1}{2}} \otimes \kappa_w \ .$$

Lets now turn to the composite operators obtained by composing $\mu^{\frac{1}{2}}Q_\Delta \nu^{-\frac{1}{2}}$ and $\mu^{\frac{1}{2}}Q_\sigma \nu^{-\frac{1}{2}}$ with (17.5), i.e. the operators:

$$(17.21) \qquad\qquad\qquad \mu^{\frac{1}{2}}Q_\Delta \nu^{-1}R_F^t \mu^{\frac{1}{2}}$$

and

$$(17.22) \qquad\qquad\qquad \mu^{\frac{1}{2}}Q_\sigma \nu^{-1}R_F^t \mu^{\frac{1}{2}} \ .$$

We know from §16 that (17.21) and (17.22) are not Fourier integral operators in the usual sense; however, their singularities are relatively innocuous: their Schwartz kernels are Lagrangian distributions on the space obtained by blowing up PxP along its diagonal, and the micro-support of these distributions is the set (16.6). In particular (17.21) is a Fourier integral operator, microlocally, on T*P-W and its micro-support is the union of the diagonal and the graph of σ. Lets compute its symbol on these two sets, starting with the diagonal: The diagonal contribution to the symbol of $g \in \Gamma$ is obtained by composing (17.8) with (17.20) which gives, formally:

$$\langle w, w' \rangle_{z'} F(z)\mu_p^{\frac{1}{2}} \otimes \nu_p^{-\frac{1}{2}} \otimes \kappa_w \otimes \mu_p^{\frac{1}{2}} \otimes \nu_p^{-\frac{1}{2}} \otimes \kappa_w$$

or in view of (17.9):

$$(17.23) \qquad F(z)H(g)\mu_p \otimes \nu_p^{-1} \otimes (\kappa_w)^2$$

The off—diagonal computation gives essentially the same expression except for the leading factor which is $\langle v,v \rangle_z$ or *zero* since v is a null-vector. Hence, *the symbol of* (17.21) *on graph* σ *is identically zero*. Returning to (17.23) the product

$$(17.24) \qquad F(z)\mu_p \otimes \nu_p^{-1} \otimes (\kappa_w)^2$$

can be interpreted as a non-zero density on T_g by Lemma 17.1. Moreover, it is clear that (17.24) depends smoothly on g, so there exists a smooth non-vanishing density, δ, on Γ whose value at g is (17.24). Thus we obtain (formally) the following expression

$$(17.25) \qquad H\delta$$

for the symbol of (17.21) on the diagonal. However, we still have to interpret (17.25) as the symbol of a pseudodifferential operator with micro-support on T*P-W. Going back to the microlocal diagram (15.4):

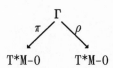

we recall that ρ maps Γ-S diffeomorphically onto T*P-W; so we can use ρ to transport the expression (17.25) to T*P-W. It is still a density, not a function; but this is easily rectified by dividing it by the Liouville density, $|\omega_P^3|$ on T*P. This gives us for the symbol of (17.21) the function

$$(17.26) \qquad\qquad \rho_* H\delta/|\omega_P^3| \ .$$

Notice by the way that if we consider this as a function on Γ-S rather than as a function on T*P-W; i.e. if we pull it back to Γ-S by using ρ again we get

$$(17.27) \qquad\qquad H\left[\delta/|\rho^*(\omega_P)|^3\right] \ .$$

By proposition 14.3 the denominator vanishes exactly to first order along S; and by lemma 17.3, H vanishes exactly to second order along S, so if g_0 is a point of S and s a local defining function for S near g_0, the expression (17.27) has to have the form

$$(17.28) \qquad\qquad |s|E$$

near g_0, where E is smooth on *all* of Γ and $E(g_0) > 0$.

Notice also that H is homogeneous of degree zero with respect to homotheties of the cotangent bundle. On the other hand, δ is homogeneous of degree *two* because of the identification, (17.3), and the presence of the κ_w^2 factor in (17.24). Thus *the total homogeneity of* (17.27) *is* -1.

To summarize, we've proved

Theorem 17.4. Microlocally on T*P-W, the operator (17.21) is the sum of a Fourier integral operator of degree -2 supported on the graph of σ and a pseudodifferential operator of degree (-1) whose leading symbol is the expression (17.26). Notice that the pull-back of this symbol to Γ is a function which is positive on Γ-S and has the form (17.28) in the vicinity of any point, $g_0 \in S$.

Similar results are true for (17.22), namely:

Theorem 17.5. Microlocally on T*P-W, the operator (17.22) is the sum of a pseudodifferential operator of degree (-2) and a Fourier integral operator of degree (-1) having σ as its underlying canonical transformation. Moreover, the pull-back of its symbol to Γ-S by the map $\sigma \circ \rho$ is everywhere positive and has the form (17.28) in the vicinity of any point $g_0 \in S$.

One can get a certain amount of information about the
on-diagonal and off-diagonal *sub-principal* symbols of (17.21)
as well, but the computations are considerably more tedious
than what we've just sketched and we won't attempt to reproduce
them here. Notice that in the construction above we are at
liberty to choose *arbitrarily* the subprincipal symbol (one term
down from the top) of Q_Δ. One can show that if it is chosen
appropriately, the subprincipal symbols of (17.21) have
singularities of the form

(17.29) $|s|E + F$

near points, $g_0 \in S$, where E and F are smooth on all of Γ.

The last stage in the construction of a parametrix for
R_F^t will involve several steps.

I.   The first step will be to "fine-tune" Q_Δ and Q_σ.
Let G be a smooth function on Γ which is homogeneous of
degree k, and let $Q_{\Delta,G}$ be an arbitrary F.I.O. with Γ as
micro-support and leading symbol equal to the product of G
with the expression, (17.19). It is easy to see that $Q_{\Delta,G}$
has properties analogous to the properties of Q_Δ listed in
theorem 17.4: For instance the composite operator

$$\mu^{\frac{1}{2}} Q_{\Delta,G} \nu^{-1} R_F^t \mu^{\frac{1}{2}}$$

is the sum of an off-diagonal term which is an F.I.O. of order
k-2, with the graph of σ as its microsupport, and an
on-diagonal term which is a pseudodifferential operator of
order k-1. Moreover, the symbol of this pseudodifferential
operator is

$$(17.30) \qquad\qquad GH\left[\delta/|\rho^*\omega_P|\right]$$

and near $g_0 \in S$ it has the form

$$(17.31) \qquad\qquad\qquad G|s|E .$$

(Compare with the expressions (17.27) and (17.28).)

We will define $Q_{\sigma,G}$ similarly: i.e. we will define
$\mu^{\frac{1}{2}}Q_{\sigma,G}\nu^{-\frac{1}{2}}$ to be any F.I.O. with Γ as microsupport and leading symbol equal to the product of G with the expression
(17.17).

II. Step 2 will be to define a kind of "$\bar\partial_b$-Laplacian" on
the manifold, P. We recall that P is a contact manifold:
To each point, $p \in P$, is attached a one-dimensional vector
space of "contact vectors," W_p in T_p^*. Let $p_0 \in P$ and U
a neighborhood of p_0. If U is small enough, there exist
vector fields, v_1 and v_2 on U with the property that for
all $p \in U$ $v_1(p)$ and $v_2(p)$ are a basis for $W_p^\perp$ in T_p.
Let Δ_U be the Laplacian, $v_2^2 + v_2^2$, on U. Now let $\{U\}$ be a

covering of P by such U's, and $\{\rho_U\}$ a partition of unity subordinate to this covering. Let

$$(17.32) \qquad \Delta_W = \tfrac{1}{2} \sum \rho_U \Delta_U + \Delta_U^t \rho_U \; .$$

By construction, the symbol of Δ_W is positive on T*P-W and vanishes exactly to second order on W. Moreover, by a theorem of Hörmander, Δ_W is subelliptic of order (-1), i.e. satisfies Sobolev estimates of the form

$$(17.33) \qquad \|\Delta_W f\|_S + \|f\|_S \geq C_S \|f\|_{S+1}, \qquad f \in C^\infty(P),$$

for all s. (See [20], part III.)

   III: The final step in the construction will be to compare Δ_W with the operator

$$(17.34) \qquad (Q_{\Delta, G} \nu^{-1} R_F^t)(Q_\Delta \nu^{-1} R_F^t) \; .$$

It is clear from theorem 17.4 that we can choose a smooth non-vanishing function, G, on Γ, homogeneous of degree 4, such that the on-diagonal component of (17.34) has the same leading symbol as Δ_W and the off-diagonal component is of order one. Morever, by (17.28) and (17.29), the symbol of the off-diagonal component and the subprincipal symbol of the on-diagonal component of (17.34) look like (17.28) near S; so there exist smooth functions, G_1 and G_2 on Γ, homogeneous of degree 3 and H_1 and H_2, homogeneous of degree two, such that if we subtract from $\Delta - (Q_{\Delta,G}\nu^{-1}R_F^t)(Q_\Delta\nu^{-1}R_F^t)$ the expression:

$$(17.35) \qquad Q_{\Delta,G_1}\nu^{-1}R_F^tQ_\Delta\nu^{-1}R_F^t + Q_{\sigma,G_1}\nu^{-1}R_F^tQ_\Delta\nu^{-1}R_F^t$$
$$+ \, Q_{\Delta,H_1}\nu^{-1}R_F^t + Q_{\sigma,H_2}\nu^{-1}R_F^t \, ,$$

the on-diagonal and off diagonal symbols of the difference are of degree zero. Let us denote this difference by P and denote by Q the sum,

$$(17.36) \quad Q_{\Delta,G}\nu^{-1}R_F^tQ_\Delta\nu^{-1} + Q_{\Delta,G_1}\nu^{-1}R_F^tQ_\Delta\nu^{-1}$$
$$+ \, Q_{\sigma,G_1}\nu^{-1}R_F^tQ_\Delta\nu^{-1} + Q_{\Delta,H_1}\nu^{-1} + Q_{\sigma,H_2}\nu^{-1} \, ,$$

and rewrite (17.35) in the form

$$(17.37) \qquad\qquad\qquad \Delta = QR_F^t + P \, .$$

To show that R_F^t satisfies an estimate similar to (17.33) (i.e. an estimate of the form (8.12)) we need Sobolev bounds of the form

$$(17.38) \qquad\qquad\qquad \|Qf\|_s \leq C_s\|f\|_{s+5/2}$$

and

$$(17.39) \qquad\qquad \|Pf\|_s \leq C_s \|f\|_s \ .$$

We will give the proof of these bounds elsewhere. They are very closely related to some estimates of Phong-Stein [33], Uhlmann [43], and Greenleaf [13], but are unfortunately not retrievable from these estimates. (In principle one can derive them from the estimates in [43] but first one has to show that these estimates are unaffected by "blowing-up".) Finally, combining (17.33), (17.38) and (17.39) we obtain for R_F^t the estimate:

$$(17.40) \qquad\qquad \|R_F^t f\|_{s+5/2} + \|f\|_s \geq C_s \|f\|_{s+1}$$

Concluding Remarks

It is possible that the material discussed in the last four sections has other applications (besides that of enabling one to implement the convergence scheme set up in §8). In particular one application which we hope to develop in more detail elsewhere is to an area which, for lack of a better term, we will provisionally call "cosmological x-ray tomo-graphy." In ordinary x-ray tomography an object in space (such as the spherical object in the figure below) is radiated by a planar family of light rays, and from the tomographic

pattern produced by these rays one tries to reconstruct the
planar slice S ∩ P.

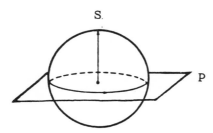

Suppose now that we blow this figure up to cosmological dimensions; that is, suppose the laboratory in which our CAT-scanner
sits is all of $\mathbb{R}^{3+1}$, and let S be a region of space-time
filled by astral debris describable by a small perturbation of
the Minkowski metric. Our problem is to reconstruct as much as
we can of the three-dimensional slice, S ∩ P, from light rays
coming to us from infinity along P. The difference between
this kind of tomography and the kind performed in the laboratory is that, in the laboratory, the distribution of mass in S
can be assumed to be *constant* and light rays can be assumed to
pass through it *instantaneously*. Modulo these assumptions the
reconstruction can be reduced to the problem of inverting the
standard (2-dimensional) Radon transform. Theoretically this
is a simple matter (though the finite difference schemes
devised by tomographers for performing this task are mind-
bogglingly ingenious). However, if the laboratory in which our

CAT—scanner sits is blown up to cosmological scale, the data
which we are trying to determine are *time-dependent*, and light
rays require *finite periods of time* to pass through the region
in which this data sits. Therefore, the problem of recon—
structing this data comes down to a problem of inverting the
relativistic x-ray transform (7.8). We have learned, however,
that this is theoretically impossible. (On mathematical
grounds it is impossible because any data which lives micro—
locally on the temporal part of $T^*M_{2,1}$ is in the kernel of
the x-ray transform. Notice, however, that there are physical
reasons for not being able to do this as well: No observer in
space-time can be privy to events occurring beyond his own
causal horizon.) Therefore, "cosmological" tomography differs
from the sort of tomography performed in laboratories in that
there are theoretical as well as practical limits to its
reconstructive capabilities.

PART V

THE FLOQUET THEORY

§18   Let (M,g) be an   n-dimensional Lorentzian manifold and let $\square$ be the conformal d'Alembertian associated with g; i.e. on a coordinate patch, $(U, x_1, \ldots, x_n)$,

$$(18.1) \qquad \square = \left[\sum |g|^{-\frac{1}{2}} \frac{\partial}{\partial x_i} g_{ij} |g|^{\frac{1}{2}} \frac{\partial}{\partial x_j} \right] + \frac{n-2}{4(n-1)} R,$$

$|g|$ being $|\det(g_{ij})|$ and   R   the scalar curvature. This operator has very simple transformation properties with respect to conformal changes of the metric, g: If Ω is an everywhere positive function and $g' = \Omega^2 g$ then

$$(18.2) \qquad \square' \Omega^{(n-2)/2} f = \Omega^{(n+2)/2} \square f .$$

This transformation law can be expressed a little more intrinsically in terms of the density bundles $|\Lambda|^{(n-2)/2n}$ and $|\Lambda|^{(n+2)/2n}$: It says that $\square$ can be regarded as a *conformally invariant* differential operator,

$$(18.3) \qquad \square : C^\infty(|\Lambda|^{(n-2)/2n}) \to C^\infty(|\Lambda|^{(n+2)/2n}) .$$

Notice by the way that if f is a section of $|\Lambda|^{(n-2)/2n}$,
then $(\square f)f$ is a section of $|\Lambda|$, so we can form the integral

$$(18.4) \qquad\qquad \int_M (\square f)f$$

(providing the support of f is compact). The expression,
(18.4), is a conformally invariant bilinear form on
$C_0^\infty(|\Lambda|^{(n-2)/2n})$; and the equation, $\square f = 0$, is just the
variational equation for (18.4).

In particular let M_0 be ordinary Minkowski space, and
let g_0 be a Lorentzian metric which is identical with the
standard Minkowski metric outside a compact set. This metric
defines a conformal structure on the conformal compactification
of M_0, i.e. on $M_{n-1,1}$; and, in particular, both (18.3) and
(18.4) are well-defined on $M_{n-1,1}$. Hence we can lift both
these objects to the universal cover of $M_{n-1,1}$, the Einstein
cylinder

$$(18.5) \qquad\qquad M = S^{n-1} \times \mathbb{R} \; .$$

In order to describe the analytic properties of these lifted
objects, we will say a few words about how the spaces, M_0 and
M, are related to each other. The fundamental group of $M_{n-1,1}$
acts on M by the deck transformation

(18.6) $\sigma:\ S^{n-1} \times \mathbb{R} \rightarrow S^{n-1} \times \mathbb{R},$ $(x,t) \rightarrow (-x, t+\pi)$

and its iterates. For $n = 2$, the Einstein cylinder is the
shaded region in the figure below

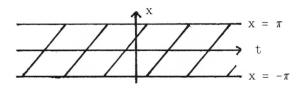

with its edges identified by the map $(t,-\pi) \rightarrow (t,\pi)$. A fundamental domain for the action which we've just described is the
region

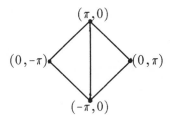

Its boundary consists of two cones, C_- and C_+, joined along
the line $t = 0$. (In three dimensions the analogous picture is

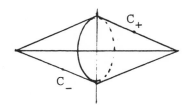

the intersection of the two cones being the unit circle in the
t = 0 plane.) σ maps C_- bijectively onto C_+; and hence
C_- and C_+ correspond to the single cone at ∞ in $M_{n-1,1}$
(see §2). As for M_0, it can be thought of as sitting inside
M as the open region bounded by C_- and C_+, C_- being its
"minus—null-infinity" and C_+ its "plus-null-infinity."

Now let f be a solution of (18.1) on $\mathbb{R}^{n-1,1}$ whose
Cauchy data are compactly supported at t = 0. If we think of
f as a $(n-2)/2n$ density, i.e. as a solution of (18.3) on the
region in the figure above (bounded by C_- and C_+), then it
is easy to see that it extends smoothly to the boundary of this
region. (In fact it extends smoothly to the whole Einstein
cylinder.) In particular, it has well defined boundary values,
which are sections, f_+ and f_-, of $|\Lambda|^{(n-2)/2n}$ over C_+
and C_-. Moreover, given the identification, $\sigma: C_- \to C_+$, we
can pull f_+ back to C_- and think of it as a section over C_-.
This gives us the so-called "scattering representation" of f:

$$(18.7) \qquad\qquad f_- \to f \to f_+ \to \sigma^* f_+ \ .$$

which describes how the "free data," f_-, at minus-null-
infinity in $\mathbb{R}^{n-1,1}$ is scattered into the "free data," $\sigma^* f_+$,
at plus—infinity. To describe this process a little more

systematically, let S be the space of solutions of (18.3) on
M. We can parametrize this space by prescribing, for each
f ∈ S its Goursat data, f_ = f ⌉ C_ . Then, in terms of this
parametrization, the scattering transformation, (18.7), *is just
the canonical action of the deck transformation*, (18.6), *on* S.

This formulation of relativistic scattering is due to
Segal and Paneitz. As Segal points out in [37], it makes sense
for lots of other operators besides (18.3). (For example, the
non-linear operator associated with $:\phi:^4$ field theory:
$\Box\phi + \phi^3$.) We refer to [37] for further detail. (See also [30]
and [31].)

We will be interested in using these ideas to define
conformal invariants for Zollfrei metrics on $M_{2,1}$ and its
double cover. Since we will be interested in the double cover
as much as in $M_{2,1}$ itself we will define our scattering
operator, not in terms of the action of (18.6) on S, but,
instead, in terms of the action of its *square*, the deck
transformation,

(18.8) $\tau: S^2 \times \mathbb{R} \to S^2 \times \mathbb{R}$, $(x,t) \to (x,t+2\pi)$

The fundamental domain of τ and its iterates is the
shaded region on the cylinder below:

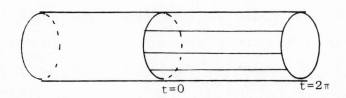

$t=0$ $t=2\pi$

and so we will parametrize solutions, f, of (18.3), in terms
of their Cauchy data on t = 0 rather than their Goursat data
on C_-.

Let g_0, for the moment, be any Lorentzian metric on
$M_{2,1}$, and let g be the metric corresponding to it on the
universal cover, $M = S^2 \times \mathbb{R}$. Before discussing the "Floquet"
aspects of this set-up we will review a few standard facts
about the conformal wave operator (18.1):

1. For each $a \in \mathbb{R}$, let M_a be the surface, t = a, in
M; i.e. $M_a = S^2 \times \{a\} \cong S^2$. Let S be the solution space of
$\square f = 0$; and, for $f \in S$, let $\|f\|_a^2$ be the total *energy*
associated with the Cauchy data of f on M_a ([20], Chapter
XXIV). Then, for all a,b, there exists (loc. cit.) a positive
constant $C = C(a,b)$ such that

(18.9) $\|f\|_a^2 \leq C \|f\|_b^2$.

2. Given $f, g \in C^\infty(M_0)$ and $h \in C^\infty(M)$ there exists a unique $u \in C^\infty(M)$ such that:

$$(18.10) \quad \square u = h \quad \text{and} \quad (u, \frac{\partial u}{\partial t}) = (f, g) \quad \text{at} \quad t = 0 .$$

Indeed, since M is spatially compact, a solution of (18.10) exists for $0 \le t \le 2\pi$. However, since $\square$ is periodic of period 2π in t one can, by iterating this solution, produce a solution of (18.10) for all time t. (loc. cit.)

3. The map

$$(18.11) \quad u \in S \to (u, \frac{\partial u}{\partial t})_{t=a} \in C^\infty(M_a) \oplus C^\infty(M_a)$$

is bijective; so we can topologize S by giving it the Frechet topology of $C^\infty(S^2) \oplus C^\infty(S^2)$. By (18.9) this topology is the same for all a.

4. There is another identification of S with $C^\infty(S^2) \oplus C^\infty(S^2)$ given by the decomposition of S into *advanced* and *retarded* solutions. This decomposition is a little more subtle than (18.11); so we will go over it in some detail: Let Σ be the characteristic variety of $\square$ (the set of *light-like* covectors in T^*M-0) and let Σ^+ and Σ^- be its two connected components. Given $p \in M_0$, there exists a

distributional function, $u_p^+ \in C^{-\infty}(M)$ such that $\Box u_p^+$ is smooth, the wave front set of u_p^+ is contained in Σ^+, and the restriction of u_p^+ to M_0 is the delta-function at p. u_p^+ is not unique; however, it is unique up to a smooth function. Moreover this smooth function can be so chosen that u_p^+ itself is smooth as a function of the parameter, p. Now let $h_p = \Box u_p^+$. By (18.10) there exists a unique smooth function, v_p with $\Box v_p = -h_p$ and v_p and $\frac{\partial}{\partial t} v_p$ equal to zero on M_0. Replacing u_p^+ by $u_p^+ + v_p$ we can assume to start with that $\Box u_p^+ = 0$.

For $p \in M_0$ and $q \in M$, let

$$u^+(q,p) = u_p^+(q) \ .$$

This distribution is the Schwartz kernel of an operator

$$(18.12) \qquad\qquad E^+ \colon C^\infty(M_0) \to S$$

called the *advanced* fundamental solution of (18.1). Notice that E^+ has been constructed so that if ι_0 is the inclusion map of M_0 into M, then

$$(18.13) \qquad\qquad \iota_0^* E^+ f = f \ .$$

Let C_0 and C be the complex conjugation operations, $C^1 f = \bar{f}$ and $Cg = \bar{g}$, on $C^\infty(M_0)$ and $C^\infty(M)$ respectively. The operator

$$(18.14) \qquad\qquad E^- = CE^+ C_0$$

is called the *retarded* fundamental solution of (18.1). Since □ is an operator with real coefficients, E^- maps $C^\infty(M_0)$ into S and satisfies the analogue of (18.13). (Notice by the way that E^+ and E^- are *not* canonically defined. Built into the definition above is a certain amount of C^∞ ambiguity.)

Combining E^+ and E^-, we get an operator

$$(18.14) \quad (f,g) \in C^\infty(S^2) \oplus C^\infty(S^2) \to u = E^+ f + E^- g \in S \ ,$$

which can, in turn, be compared with (18.11). The composite operator

$$(18.15) \qquad C^\infty(S^2) \oplus C^\infty(S^2) \to C^\infty(S^2) \oplus C^\infty(S^2)$$

can be shown to be a pseudodifferential operator. In fact its symbol, at $(x,\xi) \in T^*S^2\text{-}0$, is

$$(18.16) \qquad\qquad \begin{bmatrix} 1, & |\xi| \\ 1, & -|\xi| \end{bmatrix}$$

$|\xi|$ being the length of $\xi \in T_x^*(M_0)$ measured with respect to the Riemannian metric induced on M_0 by g. Using the index theorem one can show that the index of (18.15) is zero. Therefore, by redefining E^+ on a finite dimensional subspace of $C^\infty(M)$ one can arrange for (18.15) to be *bijective*. Having done this we obtain a decomposition of S into closed subspaces:

$$(18.17) \qquad\qquad S = S^+ \oplus S^- ,$$

S^+ being the range of E^+ and S^- the range of E^-. Furthermore, $S^- = CS^+$, and the restriction map, $\iota_0^*: S \to C^\infty(M)$, is bijective on S^+ and S^-.

Let us denote by T the Floquet action of τ (see (18.8)) on S. Since τ preserves Σ^+ and Σ^-, the Floquet decomposition of T with respect to (18.17) has the form

$$(18.18) \qquad\qquad T = \begin{bmatrix} T_{11} & T_{12} \\ T_{21} & T_{22} \end{bmatrix}$$

where T_{12} and T_{21} are smoothing, and T_{22} can be expressed in terms of T_{11} by the formula:

$$(18.19) \qquad\qquad T_{22} = CT_{11}C .$$

As for T_{11}, it is *an elliptic Fourier integral operator of order zero on* $C^\infty(S^2)$. Its microsupport is described in the figure below:

(18.20)

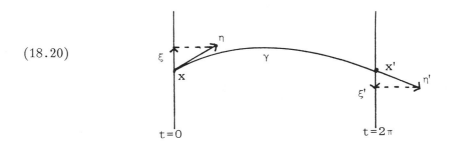

In this figure γ is a null-geodesic in T^*M-0 joining (x,η) to x',η' and ξ and ξ' are the projections of η and η' onto the cotangent space to M_0 at x and $M_{2\pi}$ at x' respectively. The map, $(x,\xi) \to (x',\xi')$, defines a canonical transformation

$$\Phi: T^*S^2-0 \to T^*S^2-0 \ ;$$

and T_{11} is a zeroth order Fourier integral operator with Φ as its underlying canonical transformation (loc. cit.).

Now consider in particular the case when g is *Zollfrei*. Then if we project the null-geodesic depicted in the figure above down onto $T^*M_{2,1}-0$, it has to close up; i.e. the image of (x,ξ) has to coincide with the image of (x',ξ'). In

other words, Φ *has to be the identity canonical transforma-*
tion. Thus T_{11} *and* T_{22} *are pseudodifferential operators.*
To compute their symbols one has to solve the transport equation for (18.1) along the curve γ in figure above. Since the
subprincipal symbol of the operator (18.1) is zero, this has
the form

$$D_{\Xi}s = 0$$

Ξ being the vector field on T^*M which defines geodesic flow.
This equation tells us that the half-density component of s
is *constant* along γ and the Maslov component *covariant
constant.* From this one deduces:

$$\sigma(T_{11}) = e^{i\sigma(\pi/2)}$$

σ being an integer modulo 4, the so-called Maslov index of γ.
(This is a $\mathbb{Z}/4\mathbb{Z}$ invariant of γ and is the *same* for all null
geodesics, γ.) Thus there exist pseudodifferential operators
of order (-1) on S^2 (say K_1 and K_2) such that

(18.21) $T_{11} = e^{(i\sigma\pi/2)}(I+K_1)$ and $T_{22} = e^{-(i\sigma\pi/2)}(I+K_2)$.

Examples

    1. Let g be the standard Einstein metric on $M_{2,1}$.
Then $\sigma = 1$ and the Floquet matrix (18.18) has the form

$$\begin{bmatrix} iI & 0 \\ 0 & -iI \end{bmatrix}$$

2. Let g be the same metric, but thought of as being on the double cover. Then the Floquet matrix is the square of the one above:

$$-\begin{bmatrix} I & 0 \\ 0 & I \end{bmatrix}.$$

From now on we will replace τ by τ^2 if necessary and assume that the Floquet operator has the form

$$\pm I + K$$

where K is a matrix of pseudodifferential operators of degree (-1). Notice that this implies that K *is compact*. Conversely suppose that g is any metric on $M_{2,1}$ or its double cover for which the Floquet operator has the form, ±I + K, where K is compact. Then, because T_{11} is an elliptic Fourier integral operator of order zero it must necessarily be a pseudodifferential operator, and Φ must be the *identity* canonical transformation. In other words *only for Zollfrei manifolds can the Floquet operator have the form, ±I + K, with* K *compact*.

The symbols of K_1 and K_2 are ordinarily rather hard to compute. We will sketch a method for computing them in

special cases in §19. There are, however, a few general re–
marks which one can make about these symbols. First of all, if

$$\sum_{r=-1}^{-\infty} k_{i,r}(x,\xi), \qquad i = 1,2,$$

is the total symbol of K_i, then, by (18.19):

$$k_{1,r}(x,-\xi) = \bar{k}_{2,r}(x,\xi) \ ,$$

for all r. On the other hand, from transmission properties of
the advanced and retarded fundamental solutions of the wave
equations (see [48]) one can easily show that

$$k_{1,r}(x,-\xi) = (-1)^r k_{2,r}(x,\xi) \ .$$

By comparing these two equations one can see that $k_{1,r}$ and
$k_{2,r}$ have to be *real* for r even and imaginary for r odd.
In particular the leading symbols of K_1 and K_2 are of the
form

$$\sigma(K_1)(x,\xi) = i\sigma(x,\xi)$$

(18.22)   and

$$\sigma(K_2)(x,\xi) = -i\sigma(x,-\xi)$$

where $\sigma(x,\xi)$ is a real-valued function on $T*S^2-0$ of degree
(-1) in ξ.

In §19 we will use the term "Floquet operator," inter-
changeably, for the operator (18.18) and for various polynomial
combinations of this operator. One Floquet operator with
particularly nice properties is the operator

$$(18.23) \qquad\qquad (T \pm I)^2$$

whose Floquet matrix has the form

$$(18.24) \qquad\qquad \begin{bmatrix} K_1^2 & 0 \\ 0 & K_2^2 \end{bmatrix} + \cdots$$

the dots indicating smoothing terms. The diagonal terms, K_1^2
and K_2^2 are pseudodifferential operators of order (-2) on S^2;
so (18.24) is an operator which is "just at the edge of trace
class," i.e. if we multiply K_1^2 by e^{-tP}, P being any positive self-adjoint elliptic operator of order one on S^2, the
expression

$$(18.25) \qquad\qquad (e^{-tP}K_1^2)/\log t$$

is of trace class for $t > 0$ and has a finite limit,

independent of P, as t → 0+. This limit is given, in fact, by the symbolic trace

$$(18.26) \qquad \int_{T*S^2} \sigma(x,\xi)^2 dx d\xi,$$

$dx d\xi$ being the symplectic volume form on $T*S^2$. (For the proof of these assertions, see [47]. Notice that the integrand in (18.26) is a homogeneous function of degree (-2) on each cotangent fiber, T^*_x. For (18.26) to make sense the integral of $\sigma(x,\xi)^2 d\xi$ over T^*_x has to be interpreted as a *residue* as in §9. (See the appendix of §9.) In particular (18.26) *is a conformal invariant of the Zollfrei metric g.*

Other conformal invariants can be generated by taking the traces of the nth powers of (18.24). Unlike (18.26) these invariants involve the *total* symbols of K_1 and K_2, and are probably much harder to compute.

A final remark: Not only is the expression (18.26) a conformal invariant of g, but the symbol, $\sigma(x,\xi)$, itself can be thought of as a conformal invariant of g. We have defined it above in an un—intrinsic way as a function on $T*S^2-0$; however we can define it more intrinsically as follows: Let Σ^+_{red} be the symplectic reduction of Σ^+ with respect to the "null-leaf" relation (the null-leaves of Σ^+ being, in this case, just the null-geodesics). Let (x,ξ) and γ be as in figure (18.20), and consider the map of $T^*S^2\text{-}0$ onto Σ^+_{red} which maps (x,ξ) to γ. This map is a symplectomorphism; and, by means of it, we can pull $\sigma(x,\xi)$ back to Σ^+_{red}. It is clear that *as a function on* Σ^+_{red}, $\sigma(x,\xi)$ *is a conformally invariant object.*

§19. The invariant (18.26) seems to behave somewhat like the
Chern-Simons invariant, [5], for compact three-folds. Namely
it seems to be an intrinsically non-local invariant whose
differential or "first variation" is given by a local recipe.
We will describe how to compute its differential in this
section. (The term "local" is a slight misnomer by the way:
We will see that the variation of (18.26) involves an
integral-differential operator in which the double fibration,
(15.2), figures in an essential way.) Our starting point is
the following observation:

Theorem 19.1. Let g_0 be a Zollfrei metric on $M_{2,1}$ and let
$\{g_t, -1 \le t \le 1\}$ be a Zollfrei deformation of g_0. Let $\square_t$ be
the conformal wave operator associated with g_t. (See (18.1).)
Then there exists an invertible zeroth order Fourier integral
operator, F_t, an invertible zeroth order pseudodifferential
operator, P_t, and a self-adjoint zeroth order pseudodifferential operator, Q_t, such that

(19.1) $$F_t \, \square_0 \, F_t^{-1} = P_t \, \square_t \, P_t + Q_t$$

Moreover, F_t, P_t and Q_t can be chosen to depend smoothly on
t and satisfy the initial conditions: $F_0 = P_0 = I$ and
$Q_0 = 0$.

Proof. We showed in §6 that there exists a canonical transformation

$$\Psi_t: \ T^*M_{2,1}{-}0 \ \to \ T^*M_{2,1}{-}0 \ ,$$

depending smoothly on t, which maps the characteristic variety
of $\sigma(\square_0)$ onto the characteristic variety of $\sigma(\square_t)$. Indeed
we constructed Ψ_t explicitly as the solution of an ODE of the
form

(19.2) $$\frac{d\Psi_t}{dt} = \Xi_{h_t} \circ \Psi_t, \qquad \Psi_0 = \text{identity},$$

h_t being a smooth function on $T^*M_{2,1}{-}0$, homogeneous of degree
one in the cotangent fiber variables. Let H_t be a first
order pseudodifferential operator with h_t as leading symbol.
Without loss of generality we can assume H_t is self-adjoint
and that its subprincipal symbol is identically zero. Let F_t
be the solution of the operator equation

(19.3) $$\frac{d}{dt}F_t = H_t F_t, \qquad F_0 = I \ .$$

By comparing (19.2) and (19.3) one can easily show that F_t is
a zeroth order Fourier integral operator, and that Ψ_t is its
underlying canonical transformation. Moreover, F_t is invertible and its inverse coincides with its formal transpose.

Let us now compute the principal and subprincipal symbols of $F_t \square_0 F_t^{-1}$. By construction, $(\Psi_t^{-1})^* \sigma(\square_0)$ vanishes on the characteristic variety of $\sigma(\square_t)$; so there exists a smooth non-vanishing function, $p_t(x,\xi)$, homogeneous of degree zero in ξ, such that

$$(19.4) \qquad\qquad (\psi_t^{-1})^* \sigma(\square_0) = P_t^2 \sigma(\square_t)$$

By Egorov's theorem the *principal* symbol of $F_t \square_0 F_t^{-1}$ is equal to the right hand side of (19.4). As for the subprincipal symbol, let

$$(19.5) \qquad\qquad \kappa_t = \sigma_{\text{sub}}(F_t \square_0 F_t^{-1}) \ .$$

Then, differentiating (19.5) with respect to t, we get

$$\dot{\kappa}_t = \sigma_{\text{sub}}([H_t, F_t \square_0 F_t^{-1}])$$

Since $\sigma_{\text{sub}}(H_t) = 0$ and $\sigma_{\text{sub}}(F_t \square_0 F_t^{-1}) = \kappa_t$, this equation can be rewritten

$$\dot{\kappa}_t = \{h_t, \kappa_t\}$$

However, $\kappa_0 = 0$; so this equation implies that $\kappa_t = 0$ for all t.

Now let P_t be a zeroth order pseudodifferential operator with leading symbol, $p_t(x,\xi)$. We can assume without loss
of generality that P_t is invertible, is self-adjoint and has
vanishing sub-principal symbol. Consider the difference

$$(19.5) \qquad\qquad Q_t = F_t \; \Box_0 \; F_t^{-1} - P_t \; \Box_t \; P_t \; .$$

The leading symbols of the two terms on the right cancel each
other out by construction, and the subprincipal symbol of the
first term is zero. As for the second term, a straightforward
symbolic computation shows that its subprincipal symbol is also
zero. Hence, the operator on the left, which at first glance
seems to be of order 2, is actually of order *zero*.   Q.E.D.

Setting

$$(19.6) \qquad\qquad V_t = -F_t^{-1} Q_t F_t$$

we can rewrite (19.1) in the form

$$(19.7) \qquad\qquad \Box_0 + V_t = F_t^{-1} P_t \; \Box_t \; P_t F_t \; .$$

This formula says, roughly speaking that, up to a conformality
factor, the wave operator associated with the metric, g_t, is
unitarily equivalent to the operator, $\Box_0$, plus a "potential."

Of course this interpretation of (19.7) has to be taken with a
grain of salt since both the conformality factor, P_t, and the
potential, V_t, are pseudodifferential operators.

Consider now the pull-backs of these operators to the
universal cover, M, of $M_{2,1}$ i.e. consider F_t, V_t, P_t etc.
as pseudodifferential operators, periodic of period 2π, on M.
By (19.7) the operator $P_t F_t$ maps the solution space of $\square_0 + V_t$
bijectively onto the solution space of $\square_t$ and intertwines the
action of the deck transformation (18.8) on these two spaces.
In particular the Floquet operator for $\square_t$ is unitarily equivalent to the Floquet operator for $\square_0 + V_t$ (which, as we will
see below, is a much more accessible object then the Floquet
operator for $\square_t$ itself.) Let us denote this operator by T_t.
By (18.18) its on-diagonal components have the form

$$(T_{11})_t = \pm(I + K_t)$$

(19.8)   and

$$(T_{22})_t = \pm(I + CK_t C)$$

where K_t is a pseudodifferential operator of order (-1). Let
$\sigma(K_t)$ be its leading symbol. To compute $\sigma(K_t)$ at $(x,\xi) \in$
$T^*S^2 - 0$ one has to solve a transport equation of the form

$$D_\Xi f_t = \mu_t$$

over the null-geodesic, γ, depicted in figure (18.20) Ξ being the bicharacteristic vector field associated with $\sigma(\square_0)$. The term on the right is of the form $\mu' + \mu''_t$ where μ' is an expression involving the first three terms in the total symbol of $\square_0$ and μ''_t is just the leading symbol of V_t. (The first term is a little complicated to write out explicitly, but fortunately it doesn't depend on t.) Integrating from zero to 2π and then differentiating with respect to t we get

$$(19.9) \qquad \sigma(\dot{K}_t)(x,\xi) = i \int_\gamma \sigma(\dot{V}_t) ds ,$$

$\sigma(\dot{V}_t)$ being the leading symbol of $\dot{V}_t$ and γ the periodic null-geodesic emanating from (x,ξ). (See figure 18.20.) The dots indicate, of course, differentiation with respect to t. Notice by the way that *even though* $\sigma(\dot{V}_t)$ *is homogeneous of degree zero in the fiber variables of* T*M-0, *the right hand side of* (19.9) *is homogeneous of degree* (-1). This is because the bicharacteristic vector field, Ξ, in (19.8) is homogeneous of degree one. Thus the parametrizing variable, s, in the integrand of (19.9) changes by a factor of $1/\lambda$ when we replace the initial point, (x,ξ), of γ by $(x,\lambda\xi)$. (Heuristically: since $\Xi = \frac{d}{ds}$ and Ξ gets changed by a factor of λ, ds gets changed by a factor of $1/\lambda$.)

Let us now examine a little more detail how the symbol of K is affected by infinitesimal cyclic deformations of $g = g_0$. As above let $\{g_t\}$ be such a deformation and let

$$\dot{g} = \left[\frac{dg}{dt}\right] (t = 0)$$

and

$$\dot{a} = \left[\frac{d}{dt} \sigma(\Box_t)\right] (t = 0) \ .$$

By definition $\dot{g}$ is a quadratic form on $T^*M_{2,1}$ and $\dot{a}$ a quadratic form on $T^*M_{2,1}$; and they are, in fact, the same quadratic form if we make the identification

$$TM_{2,1} \underset{g}{\cong} T^*M_{2,1}$$

We know from §6 that *the integral of* $\dot{a}$ *over all null bicharacteristics of* g is zero. This implies that there exists a smooth function, h, homogeneous of degree one on $T^*M_{2,1}-0$ such that

(19.10) $\quad D_{\Xi}h = \{h, \sigma(\Box)\} = \dot{a} \qquad$ when $\quad \sigma(\Box) = 0 \ .$

In fact the restriction of h to the zero set of $\sigma(\Box)$ is uniquely determined by (19.10) providing we require that it, too, have the property that its integrals over all null geodesics be zero. Moreover, as we saw in section 6,

$$\Xi_h = \left[\frac{d\Psi_t}{dt}\right](t = 0)$$

Ψ_t being the canonical transformation associated with F_t. Now let

$$H = \left[\frac{dF_t}{dt}\right](t = 0)$$

By (19.2) and (19.3),

(19.11) $h = \sigma(H)$.

Also by differentiation (19.4) and setting $t = 0$ we get

(19.12) $\dot{p} = \sigma(\dot{P}) - (\{h,\sigma(\square)\} - \dot{a})/2\sigma(\square)$.

(Notice that the expression on the right makes sense because of (19.10).) These two equations, coupled with the equations

(19.13) $\sigma_{sub}(\dot{P}) = \sigma_{sub}(H) = 0$,

completely determine $\dot{P}$ and H modulo operators of two degrees lower.

Finally lets differentiate (19.1) and set $t = 0$. This gives us the identity:

(19.14) $\dot{\square} + \dot{P}\square + \square\dot{P} - [H,\square] = -\dot{Q} = \dot{V}$.

Now let $K = \left[\dfrac{dK_t}{dt}\right](t = 0)$. By (19.9) the symbol of K at (x,ξ) is equal to i times the integral over γ of the symbol of V, or, in other words the integral of the "sub-sub-principal symbol" of the expression on the left hand side of (19.14). The "sub-sub-principal" symbol of (19.14) is not, of course, an intrinsically defined object. We claim, however, that its integral over γ is. To see this suppose we make another choice of the data, H and $\dot{P}$, in (19.11), (19.12) and (19.13). Let us denote this new data by H_1 and $\dot{P}_1$ and let $H_2 = H_1-H$ and $\dot{P}_2 = \dot{P}_1-\dot{P}$. Finally let $h_1 = \sigma(H_1)$, $h_2 = \sigma(H_2)$ $\dot{p}_1 = \sigma(\dot{P}_1)$ and $\dot{p}_2 = \sigma(\dot{P}_2)$. By (19.10) and the comment following it, $h_2 = 0$ on the zero set of $\sigma(\square)$; so there exists a smooth function, b, homogeneous of degree -1 in the cotangent variables, such that $h_2 = b\sigma(\square)$. From (19.12) and the analogous equation for h_1:

(19.15) $\dot{p}_2 = \tfrac{1}{2}\{b,\sigma(\square)\}$.

Now let B be a pseudodifferential operator of order (-1) with b as leading symbol and $\sigma_{sub}(B) = 0$. Then

(19.16) $\dot{P}_2 - \tfrac{1}{2}[B,\square]$

is of order minus-two by (19.15) and

$$(19.17) \qquad H_2 - \tfrac{1}{2}(B\square + \square B)$$

is of order minus-one. Lets denote the first of these operators by R and the second by S. Subtracting (19.14) from the
analogous equation for H_1 and $\dot{P}_1$, we get for their
difference the expression:

$$(19.18) \qquad \dot{P}_2\square + \square\dot{P}_2 - [H_2,\square] \; ,$$

which, upon making the substitutions, (19.16) and (19.17),
becomes

$$(19.19) \qquad \tfrac{1}{2}([B,\square]\square + \square[B,\square]) - \tfrac{1}{2}[B\square + \square B,\square]$$

plus the error term

$$(19.20) \qquad R\square + \square R - [S,\square] \; .$$

We leave for the reader to check that (19.19) is identically
zero; i.e. *the operator defined by (19.18) is identical with
the operator defined by (19.20)*. However, $R\square + \square R$ and $[S,\square]$
are manifestly pseudodifferential operators of order zero.

Moreover, the symbol of $R\square + \square R$ vanishes on the characteristic variety of $\sigma(\square)$ and the symbol of $[S,\square]$ is the
Poisson bracket, $\{\sigma(S),\sigma(\square)\}$, so its integral over null-
geodesics is zero. Thus the symbol of $\dot{K}$ is intrinsically
defined (independent of the choices of H and $\dot{P}$) as claimed

    To summarize what we have shown so far: let g be a
Zollfrei metric on $M_{2,1}$, let $\square$ be the conformal wave operator, (18.1), and let Σ be its characteristic variety in
$T^*M_{2,1}$. Let $\dot{g}$ be a contravariant symmetric tensor of degree
2, let $\dot{\square}$ be the first variation of the operator (18.1) with
respect to $\dot{g}$, and let $\dot{a} = \sigma(\dot{\square})$.

Theorem 19.2. Suppose that the periods of $\dot{a}$ over null
geodesics of g are zero. Then

    1. there exists a smooth function, h, on $T^*M_{2,1}-0$
which is homogeneous of degree one and satisfies equation
(19.10), viz.

$$\{h,\sigma(\square)\} = \dot{a} \ ,$$

on Σ. We can normalize h by requiring that its periods over
null geodesics of g be zero.

    2. Let

$$\dot{p} = \{h,\sigma(\square)\} - \dot{a})/2\sigma(\square) \ .$$

(See equation 19.12.) Let $\dot{P}$ and H be pseudodifferential operators on $M_{2,1}$ satisfying

$$\sigma(H) = h, \qquad \sigma(\dot{P}) = \dot{p}$$

and

$$\sigma_{sub}(H) = \sigma_{sub}(\dot{P}) = 0 .$$

Then the operator (19.14):

$$\dot{V} = \dot{\square} + \dot{P}\square + \square\dot{P} - [H,\square]$$

is a pseudodifferential operator of order zero.

    3. The integral, $\dot{\sigma}(x,\xi)$, of $i\sigma(\dot{V})$ over the null-geodesic, γ, depicted in figure (18.20) depends smoothly on x and ξ and defines a smooth function, homogeneous of degree minus-one in ξ, on T^*S^2-0.

    4. The definition of $\dot{\sigma}(x,\xi)$ depends only on $\dot{g}$; i.e. does *not* depend on the choices of h, $\dot{p}$, H and $\dot{P}$.

    5. (punch line) The first variation, $\dot{K} = (\dot{K}_t)(t = 0)$ of the Floquet operator K_t in (19.8) has leading symbol $\dot{\sigma}(x,\xi)$.

<u>Comments</u>: The map

(19.21) $\dot{g} \rightarrow \dot{\sigma}$

from the space of "cyclic" $\dot{g}$'s to the space of Floquet
symbols, $\dot{\sigma}$, can be viewed as a kind of Radon-integro-
differential operator associated with the double fibration

of §15. Of the four steps involved in defining it, all but
step 2 are relatively simple. However, in order to compute the
symbol of $\dot{V}$, in step 2, from the data, h and $\dot{p}$, one has to
make use of global symbolic computations of the kind employed
by Widom in [51]. For instance, using Widom's calculus, one
can show that the symbol of $\dot{V}$ is expressible as a differentiable polynomial in derivatives of h and $\dot{p}$ of order ≤ 3
and in the curvature tensor of g and its first covariant
derivatives. Moreover, the terms involving the curvature and
its covariant derivatives can be grouped into an expression of
the form

$$\sum R_{ijk}F_{ijk}$$

where R_{ijk} is the *Cotton tensor*. (See §11. Recall that the
Cotton tensor can be defined intrinsically as the first variation of the Chern-Simons invariant of $(M_{2,1},g)$.)

We will conclude with a few words about the status of
(19.21) as a conformal invariant of the metric g. It is clear

that, as it stands, (19.21) can't be a conformally invariant object. To begin with, in the constructions above we have treated □ as a scalar operator. We know, however, that in order for it to be conformally invariant, we have to regard it as operating on appropriate density bundles. (See (18.3).) So we have to reinstate these density bundles in the construction of F_t, P_t and V_t in (19.7). This has the effect of making the potential term, V_t, a $|\Lambda|^{2/n}$density rather than a scalar.

This, however, is not the real difficulty. The real difficulty is that the symbol, $\sigma(x,\xi)$, in (18.26) is not a conformally invariant object *as a function on* T^*S^2-0. As we explained at the end of §18, for it to be conformally invariant, one has to think of it *as a function on* Σ^+_{red}. However, if we vary the metric, g, the space Σ^+_{red} varies as well. Thus, if $\{g_t\}$ is a family of Zollfrei metrics, the symbolic data associated with $\{g_t\}$ are a family $\{\Sigma^t_{red}, \sigma^t(x,\xi)\}$ of manifold-function pairs (not just a family of functions on a fixed manifold); so the definition of $\dot\sigma$ in (19.9) is not completely intrinsic.

To see what can be said about the conformal status of (19.21), lets examine the problem which we've just described from a slightly more general perspective: Let W be a manifold and (M_t, P_t, π_t, f_t) a family of quadruples consisting of

1. a submanifold, M_t, of W,

2. fiber mapping $\pi_t: M_t \to P_t$

and

3. a smooth function, f_t, on P_t (or, alternatively, a
smooth function, f_t, on M_t which is constant along the
fibers of π_t.)

Let us assume that all this data depends smoothly on t.
In particular lets assume that the diffeotype of (M_t, P_t, π_t)
is the same for all t; i.e. let us assume that there exists a
diffeomorphism, $\Psi_t \colon M_0 \to M_t$ which maps fibers of π_0 onto
fibers of π_t, depends smoothly on t and is equal to the
identity for t = 0. Given such a diffeomorphism we can define

$$\dot{f} = \left[\frac{d}{dt} f_t \circ \Psi_t \right] (t = 0) \ .$$

Since Ψ_t is a fiber mapping, $\dot{f}$ is constant along the fibers
of π_0, and so it can be thought of as a function on P_0. The
question we have raised is: To what extent is $\dot{f}_t$ an invariant
of the deformation (M_t, P_t, π_t, f_t)? It is clear that every
diffeomorphism of M_0 onto M_t which maps fibers of π_0 onto
fibers of π_t, depends smoothly on t, and is equal to the
identity at t = 0, can be written in the form, $\Psi_t \circ \rho_t$, where
ρ_t is a diffeomorphism of M_0 onto M_0 mapping fibers of π_0
onto fibers of π_0, depending smoothly on t, and equal to the
identity for t = 0. Now consider the derivative of $f \circ \Psi_t \circ \rho_t$
at t = 0. This has the form

$$f + D_\Xi f_0$$

where $\Xi = (d\rho_t/dt)(t = 0)$. Therefore, $\dot{f}$ is not intrinsically defined on all of M_0. Notice, however, that it *is*
intrinsically defined on the critical set of f_0. In fact
since $\dot{f}$ is constant along the fibers of π_0 we can think of
it as intrinsically defined on the critical set of f_0 on P_0.
In particular, *if f_0 is constant, $\dot{f}$ is intrinsically
defined on all of P_0.*

Lets return to the problem we were looking at above: Let
(g_t) be a family of Zollfrei metrics on $M_{2,1}$ and let Σ_t, in
$T^*M_{2,1}-0$, be the characteristic variety of the wave operator
associated with g_t. We showed in §6 that there exists a symplectomorphism, $\Psi_t: T^*M_{2,1}-0 \to T^*M_{2,1}-0$ which depends
smoothly on t, is equal to the identity for $t = 0$, and maps
Σ_0 onto Σ_t. Being a symplectomorphism, it also maps the
null-fibers of Σ_0 onto the null-fibers of Σ_t; so the situation here is the situation described in the last paragraph. In
particular

Proposition 19.3. If the Floquet symbol, $\sigma_0(x,\xi)$, is identically zero, the map (19.21) is intrinsically defined in a
conformally invariant way.

For example this theorem applies to the standard Einstein
metric on $M_{2,1}$. However, for the standard Einstein metric on
$M_{2,1}$, the domain of (19.21) is the space of mass-zero, spin 2
fields and the range of (19.21) is the space of homogeneous
functions of degree (-2) on $\mathbb{R}^4$-0. $Sp(2,\mathbb{R})$ acts irreducibly
on both these spaces; and it is easy to see that these two
representations are inequivalent. (For instance their K-types
are completely different.)

Conclusion: *If g is the standard Einstein metric the
transformation (19.21) is identically zero.*

BIBLIOGRAPHY

§§1-17:

1. J. K. Beem and P. Ehrlich, "Geodesic completeness and stability", preprint, U. of Missouri.

2. M. Berger and A. Besse, *Manifolds all of whose geodesics are closed.* Springer-Verlag, NY (1978).

3. T. Branson and B. Oersted, "Conformal indices of Riemannian manifolds", *Composition Math.* (to appear).

4. R. Bryant et al, *Integral geometry*, Proceedings of a summer research conference held Aug. 12-18, 1984, vol. 63, AMS Contemporary Math. Series (1986).

5. S. S. Chern and J. Simons, "Characteristic forms and geometric invariants," *Annals of Math.* Vol. 99 (1974) 48–69.

6. Y. Colin de Verdiére, "Sur le spectre de operateurs elliptiques á bicharacteristiques toutes periodiques," *Commentarii Math. Helv.* vol. 54 (1979) 508–522.

7. M. Eastwood, R. Penrose and R. O. Wells Jr., "Cohomology and massless fields," *Comm. Math. Phys.* 78 (1980) 305–351.

8. C. Fefferman and C. R. Graham, "Conformal invariants," preprint, U. of Washington (1984).

9.   P. Funk, "Ueber Flaechen mit lauter geschlossenen
      geodaetischen Linien," *Math. Ann.* 74 (1913) 278-300.

10.   J. Gasqui and H. Goldschmidt, *Déformations infinitesmales
      des structures conformes plates.* Birkhauser, Boston
      (1984).

11.   I. M. Gelfand and M. I. Graev, "Integral transforms
      associated with complexes of lines in a complex affine
      space," *Doklady Akad. Nauk.* SSSR 138 (1961) 1266.

12.   I. M. Gelfand and G. E. Shilov, *Generalized Functions*,
      Vol. 1, Academic Press, NY (1964).

13.   A. Greenleaf, "Singular integrals with conical singu-
      larities," preprint, U. of Rochester (1985).

14.   V. Guillemin, "The Radon transform on Zoll surfaces,"
      Advances in Math. 22 (1976) 85-119.

15.   V. Guillemin and S. Sternberg, *Geometric Asymptotics.*
      Math. Surveys no. 14 AMS Providence, R.I. (1977).

16.   V. Guillemin and G. A. Uhlmann, "Oscillatory integrals
      with singular symbols," *Duke Math. J.* 48 (1981) 251-267.

17.   V. Guillemin and S. Sternberg, "Symplectic techniques in
      physics," Camb. U. Press, Cambridge (1984).

18.   S. W. Hawking and G.F.R. Ellis, *The large scale structure
      of space time.* Camb. U. Press, Cambridge (1974).

19.   S. Helgason, "The Radon transform on Euclidean spaces,
      compact two-point homogeneous spaces and Grassmann
      manifolds." *Acta Math.* 113 (1965) 153–180.

20.   L. Hörmander, *The Analysis of Linear Partial Differential Operators III and IV*, Springer-Verlag Heidelberg (1984).

21.   H. P. Jakobsen and M. Vergne, "Wave and Dirac operators, and representations of the conformal group," *J. Funct. Anal.* 24 (1977) 52-106.

21.   H. P. Jakobsen, B. Oersted, S. M. Paneitz, I. E. Segal and B. Speh, "Covariant chronogeometry and extreme distances: elementary particles," *Proc. Nat. Acad. Sci.* USA 78 (1981) 5261-5265.

22.   M. Kashiwara and M. Vergne, "On the Segal–Shale–Weil representations and harmonic polynomials," *Inventiones Math.* 44 (1978) 1-47.

23.   K. Kiyohara, "C_ℓ-metrics on spheres," *Proc. Jap. Acad.* 58, ser. A (1982) 76-78.

24.   R. S. Kulkarni and F. Raymond, "Three-dimensional Lorentz space forms and Seifert fiber spaces," preprint, U. of Michigan (1983).

25.   C. LeBrun, "H-Space with a cosmological constant," *Proceedings R. Soc. London*, A380 (1982) 171–185.

26.   R. B. Melrose and G. A. Uhlmann, "Lagrangian intersection and the Cauchy problem," *Comm. on Pure and Appl. Math.* 32 (1979) 483–519.

27.   R. B. Melrose, "The wave equation for a hypoelliptic operator with symplectic characteristics of codimension 2." *J. d'Analyse Math.* 44 (1984) 134-182.

28.   V. S. Molchanov, "Harmonic analysis on pseudo–Riemannian
      symmetric spaces of the group, SL(2,ℝ)." *Math. USSR* 118
      (1982) 493-506.

29.   B. O'Neill, *Semi-Riemannian geometry with applications to
      relativity.* Academic Press, NY (1983).

30.   S. M. Paneitz and I. E. Segal, "Analysis in space-time
      bundles. I. General considerations and the scalar
      bundle." *J. Funct. Anal.* 47 (1982) 78-142.

31.   S. M. Paneitz and I. E. Segal, "Analysis in space-time
      bundles. II. The Spinor and form bundles," *J. Funct.
      Anal.* 49 (1982) 335-414.

32.   R. Penrose, "Non-linear gravitons and curved twistor
      theory," *Gen. Relativ. Gav.* 7 (1976) 31-52.

33.   P. H. Phong and E. M. Stein, "Singular integrals related
      to the Radon transform and boundary value problems,"
      Proc. Nat. Acad. Sci. USA 80 (1983) 7697-7701.

34.   P. Scott, "The geometry of 3–manifolds," *Bull. London
      Math. Soc.* 15 (1983) 401-487.

35.   I. E. Segal, *Mathematical cosmology and extragalactic
      astronomy.* Academic Press, NY (1984).

36.   I. E. Segal, "Covariant chronogeometry and extreme
      distances. III macro-micro relations," *Int. J. Theoret.
      Phys.* 21 (1982) 851-869.

37.   I. E. Segal, "Induced bundles and non-linear wave equa-
      tions," in *Proceedings, Conference in Honor of G. W.
      Mackey*, MSRI, Berkeley (1984).

38.   I. E. Segal, "The cosmic background radiation and the
      chronometric cosmology," in *Proceedings, Rome Conference
      on the Cosmic Background Radiation and Fundamental
      Physics* (1984).

39.   C. L. Siegel, *Symplectic Geometry*, Academic Press, NY
      (1943).

40.   B. Speh, "Degenerate series representations of the uni-
      versal covering group of SU(2,2)," *J. Funct. Anal.* 33
      (1979) 95-118.

41.   S. Sternberg, *Celestial Mechanics Part II*, Benjamin, NY
      (1969).

42.   G. A. Uhlmann, "Pseudodifferential operators with double
      involutive characteristics," *Comm. in P.D.E.* Vol.' 2
      (1977) 713-779.

43.   G. A. Uhlmann, "On L^2-estimates for singular Radon
      transforms," preprint, U. of Washington (1985).

44.   D. A. Vogan, Jr., "The Kazhdan-Lusztig conjecture for
      real reductive groups" in *Representations theory of
      reductive groups*, Progress in Math. vol. 40, Birkhauser,
      Boston (1983) 223–264.

45.   A. Weinstein, "Asymptotics of eigenvalue clusters for the
       Laplacian plus a potential," *Duke Math. J.* 44 (1977)
       883-892.

46.   O. Zoll, "Ueber Flaechen mit Scharen geschossenen
       geodatischen Linien," *Math. Ann.* 57 (1903) 108-133.

Additional references for §§18-19.

47.   S. Eisen, "Traces and Hilbert-Schmidt norms for pseudo–
       differential operators in the critical Sobolev dimen–
       sion," M.I.T. thesis (1983).

48.   A. Hirschowitz and A. Piriou, "Propriétés de transmission
       pour les distributions intégrals de Fourier," *Comm.*
       Partial Differential Equations 4 (1979) 113-217.

49.   L. Hörmander, "The spectral function of an elliptic
       operator," *Acta Math.* 121, (1968) 193-218.

50.   L. Hörmander, "Fourier Integral Operators I," *Acta Math.*
       127 (1971) 79-183.

51.   H. Widom, "A complete symbolic calculus for pseudo–
       differential operators," *Bull. Sc. Math.* 104 (1980)
       19-63.